Henry Laver

The Mammals, Reptiles, and Fishes of Essex

A Contribution to the Natural History of the County

Henry Laver

The Mammals, Reptiles, and Fishes of Essex
A Contribution to the Natural History of the County

ISBN/EAN: 9783337326005

Printed in Europe, USA, Canada, Australia, Japan

Cover: Foto ©berggeist007 / pixelio.de

More available books at **www.hansebooks.com**

ESSEX FIELD CLUB SPECIAL MEMOIRS.—VOL. III.

THE MAMMALS, REPTILES, AND FISHES OF ESSEX:

A Contribution to The Natural History of the County.

BY

HENRY LAVER, M.R.C.S., F.S.A., F.L.S.,

Vice-President of the Essex Field Club, and Senior Surgeon to the Essex and Colchester Hospital, Colchester.

WITH EIGHT FULL-PAGE, AND TWO HALF-PAGE ILLUSTRATIONS.

Chelmsford:
EDMUND DURRANT & CO., HIGH STREET.

Buckhurst Hill:
THE ESSEX FIELD CLUB.

London:
SIMPKIN, MARSHALL & CO., LTD., STATIONERS' HALL COURT

1898.

CONTENTS.

	PAGES
INTRODUCTION	1
MAMMALS	31
REPTILES	84
BATRACHIANS	86
FISHES	88
ADDENDA	123
APPENDIX A. (LIST OF THE PRINCIPAL AUTHORITIES REFERRED TO)	125
APPENDIX B. (LIST OF SUBSCRIBERS)	129
INDEX	135

LIST OF ILLUSTRATIONS.

	TO FACE PAGE
HEAD OF FALLOW BUCK FROM EPPING FOREST (*Frontispiece*)	i.
BADGER EARTH IN EPPING FOREST	43
FOX EARTH IN EPPING FOREST	51
HAUNTS OF DEER IN EPPING THICKS	71
HEAD OF FALLOW BUCK FROM THE WEALD HALL HERD	75
COMMON RORQUAL, CAPTURED IN THE RIVER CROUCH, FEB. 12TH, 1891	78
VIEW OF MILL CREEK, FINGRINGHOE, ESSEX	82
ESTUARY OF THE BLACKWATER, FROM WEST MERSEA	90

ILLUSTRATIONS IN THE TEXT.

	PAGE
JACKAL FROM ONGAR WOODS	54
ANTLERS OF THE LAST RED DEER FROM HAINAULT FOREST	72

CORRIGENDA.
Page 116: For *Rophinesque* read *Rafinesque*.

PREFACE.

THE advantages of Local Lists, and the vital importance of such to all persons desiring to study the geographical distribution of animals, are so apparent that I feel no apology is needed for the publication of this Catalogue of the Vertebrate Fauna of Essex, excluding the Birds. No attempt towards the formation of such a catalogue for this county has previously been made, with the single exception of a list compiled by myself, and published in the *Transactions of the Epping Forest and County of Essex Naturalists' Field Club* (vol. ii., 1882, pp. 157-180). The list referred to, incomplete as it was, and confined to the Mammalian Fauna only, has been found of considerable use by others (if I may judge by the number of references to it), and is incorporated in the present volume.

The present work and Mr. Miller Christy's *Birds of Essex*, published by the Essex Field Club in 1891, together form a complete catalogue of the Vertebrate Fauna of the County.

It is, I suppose, inevitable that, as the pursuit of science becomes more widely spread, and investigations more exact, alterations should be made in the nomenclature and classification of the various species. Such changes, however, add immensely to the difficulty of compiling a catalogue. In the nomenclature adopted throughout these pages, I have to thank Professor Sir William Flower, Director of the British Museum, Natural History, at South Kensington, for valuable assistance with the Carnivora, Rodentia, and Ruminantia. In the orders Cetacea and Pinnepedia, I have followed the authors of Bell's *British Quadrupeds* (1874), and Mr. Thomas Southwell in his *Seals and Whales of the British Seas* (Norwich, 1881). To him also my thanks are due. The late Mr. G. E. Dobson, F.R.S., having been the recognised authority for Chiroptera and Insectivora, I have adopted the nomenclature of his *Catalogue of the Chiroptera* (1878), and of his *Monograph of the Insectivora* (1890).

I have to thank Mr. G. A. Boulenger, F.R.S., for kind help in naming the Reptilia and Amphibia in accordance with the methods adopted by the authorities of the British Museum.

Essex fishes have been so little investigated by Essex naturalists that it is certain numerous species must be omitted from a catalogue compiled, as this to a great extent is, by one observer alone. But, if we have scanty local lists to refer to, we have in Dr. Day's *Fishes of Great Britain* (1880) a nomenclature and arrangement that leaves nothing to be desired. I have adopted these throughout, and have likewise freely used the records there given of species captured in the county. For the rest, I look forward to a considerable addition to the number of Essex Fish when, if ever, a second edition of the present work is required. Should its publication arouse greater attention to and study of this important class, I shall feel amply repaid for the labour of preparing it.

A list of the principal works which have been referred to in the preparation of the present volume is given at the end in the form of an appendix.

My thanks are also due to Mr. E. A. Fitch, F.L.S., for much acceptable help; to Mr. Thomas Taylor, M.R.C.S., of Bocking, for a list of the fresh-water fish of the Blackwater and Chelmer rivers; to Mr. William Cole, F.L.S., of Buckhurst Hill, Honorary Secretary to the Essex Field Club, for many valuable suggestions made whilst the work was passing through the press; and to Mr. Walter Crouch, F.Z.S., of Wanstead, for several suggestions and for the loan of his drawing of the Rorqual captured in the river Crouch.

To Miss C. Fell Smith, I am indebted for

much valuable help in supervising and correcting the manuscript of this work for the Press, as also for her assistance during its passage through the printer's hands. Without her kind offices, I fear it would have been much more imperfect in every way.

I also owe to Mr. Miller Christy a debt of gratitude for the help he has so ungrudgingly given in preparing the manuscript of this Catalogue for the press, as well as for the trouble he has taken in making the needful business arrangements for bringing it before the public. In both these respects, his technical knowledge has been of the greatest possible service to me.

I am duly sensible of the imperfections, literary and scientific, of this work, and would ask my critics to bear in mind that it has been prepared largely from my own notes, with few opportunities for visiting the Metropolis, and in intervals snatched from the engagements and demands of a busy professional life.

HENRY LAVER.

Head Street, Colchester.
February, 1898.

THE MAMMALS, REPTILES, AND FISHES OF ESSEX.

INTRODUCTION.

ESSEX, the tenth English county in point of size, contains an area of some 987,623 acres. It is of irregular shape, and measures sixty-three miles in length, from S.W. to N.E. On the south, it is bounded by the Thames; on the north, by the river Stour; the rivers Lea and Stort divide it on the west from Hertfordshire; and its eastern limit is the North Sea.

The sea-board of the county is deeply indented by the four large estuaries of the rivers Thames, Crouch, Blackwater and Stour, upon the shores of which, as on the intervening sea-line, a wide margin of marsh occurs, more or less broken into low-lying islands and tracts connected with the mainland.

The shores of this marshy district consist of long stretches of sand or mud, exposed at low water, and intersected by creeks, many of them also dry at low tide. The sea surrounding the Essex coast is everywhere shallow, with a muddy or sandy bottom, not a particle of hard rock being exposed in any single part. The conditions of the sea-board, therefore, while being most favourable for certain genera of fish, are entirely unsuited to others. The rock-loving species, for instance, are either very rare or totally absent. The same may be said of the fish that live in deep water.

Inland, Essex consists of an elevated table-land, fairly timbered, and well cultivated in all parts. With the exception of the Waltham Forest district, the county is now without any extensive woods or large wastes, although these existed in former times. Large sheets of inland waters are also absent; and, owing to the level formation of the surface, all the rivers have a slow current. Although the valleys are at times particularly subject to floods, yet the valley-soil is not boggy, the land being firm up to the very edges of the streams.

The climate of Essex is dry, the average rainfall being lower than in any other English county.

MAMMALS.

The highly-cultivated condition of the county has been unfavourable to the continued existence of several of our larger Mammals, although the Epping

Forest district has been a haven of refuge for some that would otherwise have been extinct.

Yet, with the total number of recognised British Mammals—seventy-two, viz., forty-five terrestrial and twenty-seven marine—the list of Essex Fauna compares very favourably.

Of the former, thirty-eight (not counting two doubtful Bats, *Rhinolphus ferrum-equinum* and *Vespertilio murinus*, which I have thought fit to reject) will be found described in the following pages, while another, an introduced animal (a species of Jackal), probably exists in Epping Forest.

It is true that three of these species, the Badger, Marten, and Polecat, are now rare, especially the last two; yet evidence satisfactorily shows that in the early part of this century all of them were fairly abundant, and up to the present time individuals have since continuously existed.

Deer, in a wild condition, exist to-day in very few English counties. Yet, in consequence of the survival in Essex of the virgin woods of Epping Forest, we are enabled to claim these interesting animals as members of our Fauna, as they undoubtedly have been from time immemorial. Fallow Deer remain until the present time. Red Deer were known up to the early years of this century, when the presumed last surviving members of the wild herd were removed to Windsor, but stags lingered in the Forest at least as late as 1827 (See *Proc. E. F. C.*, vol. i., xlvii.) A few Red Deer have since been re-transferred to the Forest, in the hope of restoring the original stock, but they proved so destructive to the crops of neighbouring farmers that they had to be destroyed. The Roe Deer, which appears to have been by no means rare in late mediæval times, became extinct for many years; but this species also has been re-introduced, from Dorsetshire, and is doing well. We hope that, by the restoration of the Red Deer, we may soon be able to say that all the species of British Deer exist in a wild condition in Essex.

Under the head of terrestrial Mammals, it may be pointed out that Essex, as compared with many counties, shows a high record of Bats, eight out of the sixteen species given by Bell having been actually found in the county, without including the two others named by Bell and Newman, which, as will be afterwards explained (see pp. 31 and 123), must not be placed to the credit of Essex. This abundance of Chiroptera may be in part owing to more careful observation of that particular Order, but it is also owing to the fact that our county is within the northern limits of several species. Mr. Christy's records of the capture of two specimens of the Serotine seem a proof that still further attention would result in more finds, for it is difficult to believe that the only two stragglers of this species should have come under one observer's notice. An interesting paper on the Bats of Epping Forest, by the late Edward Newman, F.L.S., is reprinted in *Essex Naturalist* (vol. ix., pp. 134-138).

No reference will be made in this list to extinct British Mammals. The number of these which our present conditions of climate would suit is very small, and investigations concerning them, however interesting, give little help in determining the geographical distribution of the Fauna of to-day. It may be mentioned, however, in passing, that the only extinct British Mammals of whose existence in Essex we have any direct proof, are the Wild Swine, the Wolf, and possibly, the Beaver.

The following interesting historical reference to the first-named animal, the Wild Boar, was contributed by Mr. Charles K. Probert, of Newport, Essex, to *The Essex Note-Book and Suffolk Gleaner* (Colchester, Oct. and Nov., 1885, p. 136). His information was derived from a manuscript among the muniments of Colne Priory, which contains the following passage:

"The Survaye of the Lordshipps and Manors of Erles Colne and Colne Priory, parcel of the possessions of Roger Harlakenden, Esq., made in Anno Dni. 1598, by Israel Amyse, Esq.

"Chalkny Wood contcineth 168a. 3r. 28p. This

Wood in tymes paste was empaled; and the Erles of Oxenforde in former tymes (for their pleasure) bredd and maintayned wilde Swyne in the same untill the Reigne of King Henry the Eight. About wch tyme they were destroied by John then Erle of Oxenford, for that he understode that the Inhabitants therabout sustained by them very great Losse and Damage."

Neither is it intended to give here any lengthy notices of the domestic animals in the county. Yet, since, in other parts of the kingdom, there exist peculiar breeds of such antiquity as almost to appear indigenous, it may be well to explain briefly why we have none here. Devoid as we are of special districts requiring special varieties of domestic animals, these have not been produced in answer to a want which has never existed. It is true that some of these peculiar breeds have been so altered by crossing that, in the endeavour to meet later methods of agriculture, they have lost their original characters; but the greater number of local breeds are so suited to the individual conditions of their district that it has not been found either desirable or profitable to exchange them for any other, even when a given breed has elsewhere been found more economical or useful. For instance, the generally-useful Shorthorn Ox, with its valuable capacity for early maturity, can never replace the Black Cattle of Pembroke, nor those of the mountains of North Wales or Scotland. Neither can the South Down Sheep, with all their excellence, supplant either those of the mountainous parts of the kingdom or of the marshes of Romney. Youatt says: "In all the different districts of the kingdom, we find various breeds of sheep [and, he might have added, cattle] beautifully adapted to the locality which they occupy. No one knows their origin: they are indigenous to the soil, climate, and pasturage, the locality on which they graze: they seem to have been formed by it and for it." This is very true, and it would seem that less has been done by the agriculturist than by nature. In consequence of the large growth of cereals in Essex, and the absence of hills and wastes, it has not been found profitable to keep, for any great number of years, any strain of animals in any particular district. The result has been, therefore, that natural effects and environments have not had sufficient time to produce those modifications in the race which result in the establishment of a localised race or breed. It is to the constant changes in the numbers of the stock on our corn lands, and to insufficient facilities for rearing young stock, less than to the lack of skill and enterprise amongst our agriculturists, that we may attribute the non-localisation of a special breed of cattle

The large estuaries of Essex, to which allusion has

already been made, form traps for the capture of a considerable number of Marine Mammals, which from time to time are stranded on their shores or become entangled in their numerous sand-banks. The list of these accidental visitors (for such, as a rule, they are) is, proportionately, much above the average of most of the maritime counties of England, where the same facilities for their capture do not exist.

Ten Marine Mammals, out of the total of twenty-seven, have been noted in Essex. These include the Porpoise and the Bottle-nosed Dolphin, both of which may be frequently seen off this coast, and may therefore count as regular, not accidental, visitors

REPTILIA AND AMPHIBIA.

Essex has no extensive sandy wastes to form a home for such British Reptiles as are found flourishing on similar spots in Hampshire and some other parts of England. Nevertheless, our list of Reptilia is not far behind that of other counties, and comprises four out of the nine species on the British lists. Enclosure and cultivation are having an unfavourable effect on the numbers of those species which still remain to us; and, in some districts, indeed, it would seem as if the Snakes were becoming extinct and would altogether disappear before many years are past. This, as naturalists, we must altogether regret, although the public generally will perhaps consider it rather a matter for congratulation.

If it be a fact that the Sand Lizard occurs in Northumberland and Durham, it is very remarkable to find it absent from counties much farther south of them. True, there have been several reported cases of its capture in Essex, but in every instance a mistake has been proved in the identification. The probability therefore is that it does not occur here. The same may be said of the Smooth Snake, which is also absent from this county.

For Amphibians, we are better off. The British lists include seven species, of which we have six. One of these, the Edible Frog, is an introduction, as it probably is wherever found in the United Kingdom. The other Amphibian wanting is the Natterjack Toad, which is found both in Suffolk and Norfolk. There being no conditions unfavourable to it in Essex, it may possibly exist, although as yet unobserved.

FISH.

Turning now to the class Pisces, we find that, out of the 234 distinct species given by Francis Day (*Fishes of Great Britain and Ireland*) for these islands, 113 have been already observed in Essex. This list, compiled as it is chiefly from the observations of a single naturalist, unassisted save by a few scattered records in local papers, and these without any exactitude of detail, can by no means claim to be final.

It might have been expected that, in at least one of our numerous fishing villages, there would have been found some educated and intelligent observer who would interest himself in the study of how his fishing neighbours subsisted, on what their marketable fish fed, and what species were brought in their nets to the surface. Unfortunately, however, this branch of Natural History appears to have been entirely neglected in this district, and those who have cultivated scientific tastes have apparently confined them to the more generally interesting studies of Ornithology, Entomology, and especially Lepidoptera.

It may be that the great difficulties in preservation and the impossibility of retaining the splendid piscal colouring have contributed to this result. Anyhow, whatever the cause, the fact remains that there are few Essex records, and no later lists to refer to than that in Dale's *History of Harwich and Dovercourt* (1732) and those of later writers who have copied from him. Considering the class from which our professional sea-fishermen are drawn, it is hardly to be expected that any records by them would find their way into print. From my knowledge of them, I should say they pay little attention to the produce drawn from their nets, except the marketable kinds. Even of these, no attempt to distinguish the different species is made: one name (and that, more often than not, a purely local one) is often applied in a most hap-hazard manner to several allied species.

Having thus pointed out some of the main difficulties confronted in compiling this portion of the list, I may add that I have found it necessary to give only those species of fish which have been either identified by myself, or mentioned by some authority, such as Yarrell, Day, and Donovan, or noted by some other competent recorder in the pages of *The Zoologist*, *The Field*, or *Land and Water*.

The lack of catalogues of the Fauna of the maritime counties generally has been a still further hindrance. A list of the Fish Fauna of all our neighbouring counties would afford opportunities for comparison, and, by teaching us what species exist on similar shores, would incite a search for those whose presence would be otherwise unsuspected.

Of the three counties on the eastern coast of England for which lists have been published (*viz.*, Norfolk, Yorkshire, and Essex), the former has been especially fortunate. Sir Thomas Browne (1605-1682), the well-known author of the *Religio Medici*, compiled, more than two centuries ago, "An Account of Fishes found in Norfolk and on the Coast." This was first printed by Simon Wilkin in his edition of Browne's

Works (London and Norwich, 1835, 8vo, vol. iv., p. 325). Charles John and James Paget published *A Sketch of the Natural History of Yarmouth* (Yarmouth, 1834, 8vo); the Rev. Richard Lubbock included in his widely-known *Observations on the Fauna of Norfolk* (Norwich, 1845, 8vo) a list of the fresh-water fish of the rivers and broads; and Dr. John Lowe has published *(Trans. Norf. and Norwich Nat. Society,* 1874, p. 21) a "List of Norfolk Fish," which included 143 species.

Messrs. W. Eagle Clarke and W. D. Roebuck's list of sea and river fish of Yorkshire, given in their *Handbook of the Vertebrate Fauna of Yorkshire* (London, 1881), comprises 148 species.

The only existing lists of the sea-fish of Essex are those given by Dale in his *History of Harwich* (2nd Ed., 1732), and by W. H. Lindsey (*A Season at Harwich,* 1851), who mainly copies Dale.

In the present catalogue, I propose, as stated, to insert 113 species. This, in comparison with the larger county, seems a fair proportion, especially when it is remembered that Yorkshire has, as well as a muddy and sandy shore, a considerable stretch of rocky coast. Moreover, Messrs. Clarke and Roebuck have included in their total some species considered by Dr. Day to be varieties only, as well as all those taken on the Doggerbank. In the present volume, only those taken actually upon the coast, or within or on those sands lying in the immediate vicinity of the coast, are enumerated.*

FISHERIES.

No account of the Fish Fauna of a county can be satisfactory without reference to the various seats of its Fisheries, the methods pursued by the fishermen, and the species captured or sought in the different districts. In enumerating these, I purpose to begin with what is probably the best known, as well as the oldest—shrimping. The fishermen of Leigh have long carried on an extremely active and remunerative trade in shrimps, for which they trawl in the mouth of the Thames. Of late years, however, they have been forced to go much farther out to sea than they were wont to do, in consequence of the great increase of impurities in the water, and the disturbance caused by the passing up and down of large numbers of steamships bound to and from the Port of London. Another section of the Leigh fishermen spends many weeks and months trawling for fish in the North Sea, forwarding the produce of their toils from the various

---

* The Committee appointed by the British Association "for the purpose of considering the question of accurately defining the term 'British,' as applied to the marine fauna and flora of our islands" have reported that the "British Marine Area" may be conveniently subdivided into a shallow-water and a deep-water district. The 100-fathom contour line is a natural boundary line for the former off the north and west coasts of the British Islands; on the south and east the only boundary is the half-way line between Great Britain and the Continent; this should include the Dogger Bank. The above district may be termed the "British Marine Shallow-water District." The "Deep-water District" of 100 to 1,000 fathoms only occurs off the north and west coasts, and consequently does not concern our Essex recorders.

ports nearest to their trawling grounds. Others, again, of late years, have embarked largely on Whitebait fishing. This tiny fish is taken either by means of a small stow-boat net or a ground-seine. The shores of the estuaries of the Thames, Crouch, and Roach are favourite localities for this trade, because they are within easy reach of the great market of London; for Whitebait is a fish that bears carriage badly, and soon loses its freshness.

The town of Barking was, at one time, the port from which the North Sea trawlers started; but, since railways have enabled fish to be sent in a fresh condition from almost any port, the importance of Barking as a trawling centre has materially declined.

Harwich has always had a large number of boats engaged in the North Sea fishery, and their number is not decreasing. Here, in 1712, was first invented the well-boat, which, in pre-railway days, enabled fish to be delivered in London in a good and sometimes almost in a living condition. The well-boat has a portion of its bottom perforated by a large number of small holes, thus letting the sea-water freely into its interior. This part is called the "well," and it is divided from the spaces at both ends, by which the boat floats, by water-tight bulk-heads. The fish, when caught, are placed in this well, where they will keep alive for a long time, as the movements of the boat afford them a constant change of water. Since the use of steam carriers and ice, these well-boats have not been of such vital importance, although they are still valuable adjuncts to the Doggerbank fishery. In addition to the deliveries of fish by boats belonging to the port, Harwich is visited by many smacks from other districts, which here land their catch for carriage to London. The safety of the harbour, the easy access to it, the shortness of the railway journey, and the facilities afforded for placing cargoes on the rail, make this port a most advantageous one.

Burnham and the Crouch supply a small number of fishermen both for in-shore and deep-sea trawling. The number is probably increasing.

Maldon, Mersea, Tollesbury, and the villages on the shores of the Blackwater have a considerable population engaged in the sea fisheries, mostly on trawlers belonging to other ports, both of the North Sea and the English Channel. I do not think any of their boats take part in the drift-net fisheries of the North Sea.

Brightlingsea, Wyvenhoe, and the shores of the Colne have a larger population engaged in fishing than any other part of the county. Their boats are engaged wherever there is fish to be caught in the seas surrounding the United Kingdom, from the western shores of Ireland, where the mackerel fishing attracts them, to the eastern parts of the North Sea, where they are occupied with trawling and dredging for oysters on the so-called "skilling grounds."

A form of trawling pursued on the Essex coast, and by no means common elsewhere, is the trawling for eels on the shores near the mouth of the rivers, and on the mud-banks of the coast just outside the rivers. Sometimes this is a very paying business; and, from the naturalist's or ichthyologist's point of view, there can be no more edifying sight than to be a spectator of the turning out of the haul, with leisure to examine the mass of mud, weeds, and living freight brought up by the trawl.

Stow-boat fishing for Sprats* is carried on in this district to a greater extent than anywhere on the East Coast, and a good Sprat season is of as great importance to the shore population as a good harvest inland is to the country generally. It means abundance and comfort to the fishermen, their wives, and families, instead of poverty and want if the season be bad. Enormous quantities of Sprats are captured in a good season, and sent away to the populous districts for food. Much larger quantities are disposed of to the neighbouring farmers for manure.

---

* The term "Stow-boating," or "Stow-netting," simply means that the "Stow-net" is being used. See Yarrell's *British Fishes* (vol. ii., p. 123) and Day's *Fishes of Great Britain and Ireland* (vol. i, p. xcix.), where an illustration of a stow-net is given.

Seine-netting is rarely used on our coast, except for Garfish, Whitebait, or Smelts.

A form of fishing practised principally on the shores of the Blackwater, by which large numbers of Codling, Mullett, and other fish are sometimes caught, is known as "petering," or "peter-netting." A peter-net consists of a net about twenty fathoms long and ten feet wide, with corks on the head-rope and leads on the ground rope. The head-rope is brought over the ground-rope by folding the net longitudinally, and then fixing it there by lashings about three feet long. To use the peter-net, a boat is rowed at high tide to some suitable place on the shore. One end of the net is then cast over in from six to thirty feet of water, having an anchor attached, and a buoy to mark the spot. The boat is rowed parallel to the shore, and the net is payed out so that its open part is towards the beach. When the end is reached another anchor and buoy is attached and thrown over, as at the opposite extremity. The fishermen now row about between the net and the shore, trailing a piece of chain or something of that kind, in order to startle the fish, who, in their endeavours to reach deeper water, run into the net. When sufficient splashing about has been made, one end of the net is drawn up, and traction being made on the ground- and head-ropes, the mouth is closed. The captive fish are then taken out as the net is under-run and taken aboard, when the same process is repeated in another place. I have dwelt rather at length on this form of fishing because I have not seen it employed elsewhere.

Kettle, or keddell, fishing is only adopted to a limited extent on the sandy shores of Foulness Island and Shoebury. At one time, this was a very successful plan, employed principally for the capture of the various species of flat-fish which frequent the shallow waters covering the sands at high tide. It resulted in a fair number of Turbot being taken, although the hauls were not confined exclusively to flat-fish, since any fish passing between the entrances of the nets was almost certainly retained.

Kettles, which may be considered as fixed seines, vary in shape, size, and plan in the different districts of the kingdom where they are in use. In Day's *Fishes of Great Britain and Ireland* (vol, i., p. ciii.), there is an illustration of the kind in use for catching Mackerel on the South Coast of England. In Foulness, kettle-nets take the form of the letter V, and are either set singly or two or more in a line, with the apex of the V, which is furnished with a purse, pointing away from the shore. These nets are about 120 yards long by 4 feet high. They are fixed in position by stakes driven into the ground; to these, both the head- and ground-ropes are attached. Thus a wide area, shallowest towards the shore, is enclosed. As the fish follow the rising tide, they are carried between the nets, and arrive in the hollow of the V; but, in their return on the falling tide, they are inevitably carried into the purse at the apex. The nets are visited when the tide is falling, and the fish are removed before they are damaged by exposure out of the water.

There are, inside the sea-walls of the island of Foulness, some shallow pits filled with sea-water, which formerly were in use for keeping the catch alive until there should be a convenient occasion for sending it to market. The more valuable fish (Turbot, for instance) had a string with a cork at the other end fastened round their tails, so that the fisherman who caught them out of these stewes, when the time came for dispatching them to market, knew exactly where to dip his large landing net, and could draw them up without risk of injury.

Kettle-netting would, no doubt, be successful on all our coasts, were it not for the large quantities of *Zostera marina* floating about, blocking the nets and, by its weight, sometimes overturning them. A further difficulty is that large portions of the shore are too soft and muddy for the men who are engaged in watching and attending to the nets to move about on while emptying, clearing, and setting them up.

It is much to be regretted that all these forms of in-shore fishing, whether for Shrimps, or Eels, or other fish, destroy an enormous quantity of the fry of valuable species, such as Turbot, Brill, Soles, and Plaice. It remains an open question whether the catching of Eels, Shrimps, etc., by small-meshed nets does not indirectly destroy fish of much greater value, which would produce a far better return to the fishermen if allowed to come to maturity. In my opinion, even if the Fishery Board (by determining the minimum size of the mesh to be allowed in the district) destroyed the Whitebait fishery altogether, they would inflict no injury upon, but rather deserve the thanks of, the community of Essex fishermen.

RIVER FISHERIES.

The River Fisheries of Essex are of small value. The poisoning of the Thames by sewage has destroyed our only Salmon-river. Some individuals, as will be seen later on, occasionally attempt to pass up it; but, so far as I know, they never reach that small portion of the Thames in Essex which is, or rather should be, fresh water, but which, instead, only resembles the vilest and most unsavoury sewage.

The want of records of species for Essex rivers has caused much extra labour in compiling the following lists of the fish inhabiting each. No pains have been spared, however, to make them complete. Fishermen have been employed for every river, to ransack its waters by hooks and nets for possible treasures, and to send the results of their labours for examination and identification.

Fish, as a rule, in any stage of their existence, are easily identified, except in some genera, where the alteration before natural growth is complete is very great and the immature fish is so unlike its parents that another specific name has been actually conferred upon it; while sometimes, even, the immature fish has been placed under another generic name.

Yet the difficulty that many experienced fishermen have in correctly naming their captures shows how needful it is to be careful in this respect. The resemblance of allied species, such as the Rudd and the Roach, cause frequent mistakes. This similarity is so strong that I rarely meet with even an experienced angler who can give the distinguishing characters of these common fish. Judging by the notices in the various journals, few among our educated fishermen even appear to differentiate the various species they capture. It may constantly be seen recorded that so many Bream or Roach have been taken—records perfectly useless to the naturalist, and, from their imperfect nature, of little apparent value to the angler or sportsman.

I have been careful not to insert in the lists below any doubtful species, unless personal identification has satisfied me that it is an inhabitant of the river under which it is named, and, therefore, a member of the Essex Fish Fauna.

Of the small fresh-water section of England's noblest river—the Thames—which we can claim as belonging to this county, the list must necessarily be brief, owing to the impure condition of the water. From this list, scanty as it is, I fear, in truth, that some even now may be absent. Whenever that golden time arrives (as it must some day), when the Thames waters are once more pure and undefiled, the lordly salmon will again be seen passing up to its spawning grounds, while numerous other species will people the clear waters of the river, as in the days when, for convenience, fishing-tackle makers were fain to establish themselves hard by it in Crooked Lane, London.

Fish found in the River Thames.

1. Flounder, *Pleuronectes flesus.*
2. Salmon, *Salmo salar.*
3. Sea-trout, *Salmo trutta.*
4. Smelt, *Osmerus eperlanus.*
5. Allis Shad, *Clupea alosa.*
6. Twait Shad, *Clupea finta.*
7. Eel, *Anguilla vulgaris.*
8. Sturgeon, *Acipenser sturio.*
9. Lamprey, *Petromyzon fluviatilis.*
10. Sea-lamprey, *Petromyzon marinus.*

Under the Lea, the largest tributary of the Thames in our district, I shall include the Stort and its other feeders. For this river, there are many authorities, from Izaac Walton down to the later lists by Lieut. R. B. Croft, R.N., published in the *Transactions of the Hertfordshire Natural History Society* (August, 1882).

From the last-named authority, we learn that in the Lea and its tributaries may be found a number of species greater than that inhabiting any other river in the county. It may be that its proximity to the metropolis has induced enthusiastic anglers, with whom it is a favourite hunting ground, to introduce species not occurring naturally. In the paper quoted, four species—the Barbel, Chubb, Ruff, and Bleak—are mentioned, which I have not taken in any other Essex river. The Barbel is, however, to be found in the lake at Dagenham, and the Chubb has lately been introduced into the Blackwater.

Lieut. Croft, quoting from Chauncey's *History of Hertfordshire* (1700) and also from Farmer's *History of Waltham Abbey* (1735), mentions that the Salmon, formerly abundant in this river, probably became extinct there before the end of last century.* Under these circumstances, this species must still be included

---

* The Salmon was, however, certainly taken in the River Lea well into the present century (see p. 103).

in the list for the Lea. Lieut. Croft's list is as follows :

FISHES OF THE RIVER LEA.

1. Perch, *Perca fluviatilis.*
2. Ruff, *Acerina vulgaris.*
3. Miller's Thumb, *Cottus gobio.*
4. Stickleback, *Gasterosteus aculeatus.*
5. Ten-spined Stickleback, *Gasterosteus pungitius.*
6. Flounder, *Pleuronectes flesus.*
7. Salmon, *Salmo salar.*
8. Sea-trout, *Salmo trutta.*
9. Trout, *Salmo fario.*
10. Grayling, *Thymallus vulgaris.*
11. Pike, *Esox lucius.*
12. Carp, *Cyprinus carpio.*
13. Barbel, *Barbus vulgaris.*
14. Gudgeon, *Gobio fluviatilis.*
15. Roach, *Leuciscus rutilus.*
16. Rudd, *Leuciscus erythrophthalmus.*
17. Chubb, *Leuciscus cephalus.*
18. Dace, *Leuciscus vulgaris.*
19. Minnow, *Leuciscus phoxinus.*
20. Tench, *Tinca vulgaris.*
21. Bream, *Abramis brama.*
22. Bream-flat, *Abramis blicca.*
23. Loach, *Nemacheilus barbatula.*
24. Eel, *Anguilla vulgaris.*
25. Sturgeon, *Acipenser sturio.*
26. Lamprey, *Petromyzon fluviatilis.*

The Roding, the next tributary of the Thames, is, apparently, not a favourite with anglers. I cannot find any list of the fishes of this river. I have seen no species, and it probably does not harbour any that are not found in the larger rivers of the county.

The next feeder of the Thames, the Rom, must be passed over with the like remarks.

Of the Crouch, but a small extent of its course is fresh water, and the fishing in this portion is unimportant. It contains no species to add to our Fish Fauna.

The Chelmer and Blackwater are more important water-courses. The following lists have been kindly prepared for me by Mr. Thomas Taylor, M.R.C.S., of Bocking. They give us no additions to the list of Essex Fish.

FISHES OF THE RIVER CHELMER:

1. Perch, *Perca fluviatilis.*
2. Miller's Thumb, *Cottus gobio.*
3. Stickleback, *Gasterosteus aculeatus.*
4. Ten-spined Stickleback, *Gasterosteus pungitius.*
5. Trout, *Salmo fario.*
6. Pike, *Esox lucius.*
7. Carp, *Cyprinus carpio.*
8. Gudgeon, *Gobio fluviatilis.*
9. Roach, *Leuciscus rutilis.*
10. Dace, *Leuciscus vulgaris.*
11. Minnow, *Leuciscus phoxinus.*
12. Tench, *Tinca vulgaris.*
13. Bream, *Abramis brama.*
14. Loach, *Nemacheilus barbatula.*
15. Eel, *Anguilla vulgaris.*

Fishes of the River Blackwater:

1. Perch, *Perca fluviatilis*.
2. Miller's Thumb, *Cottus gobio*.
3. Stickleback, *Gasterosteus aculeatus*.
4. Ten-spined Stickleback, *Gasterosteus pungitius*.
5. Salmon, *Salmo salar*.
6. Trout, *Salmo fario*.
7. Sea-trout, *Salmo trutta*.
8. Pike, *Esox lucius*.
9. Gudgeon, *Gobio fluviatilis*.
10. Roach, *Leuciscus rutilis*.
11. Chubb, *Leuciscus cephalus*.
12. Dace, *Leuciscus vulgaris*.
13. Minnow, *Leuciscus phoxinus*.
14. Tench, *Tinca vulgaris*.
15. Loach, *Nemacheilus barbatula*.
16. Eel, *Anguilla vulgaris*.

In the Colne river, the fishing is somewhat cared for. I am able to give the complete list from my own observation. No species occurs, except Planer's Lamprey, which is not found in the Lea.

Fishes of the River Colne.

1. Perch, *Perca fluviatilis*.
2. Miller's Thumb, *Cottus gobio*.
3. Stickleback, *Gasterosteus aculeatus*.
4. Ten-spined Stickleback, *Gasterosteus pungitius*.
5. Sea-trout, *Salmo trutta*.
6. Trout, *Salmo fario*.
7. Pike, *Esox lucius*.
8. Carp, *Cyprinus carpio*.
9. Gudgeon, *Gobio fluviatilis*.
10. Roach, *Leuciscus rutilis*.

11. Rudd, *Leuciscus erythropthalmus.*
12. Minnow, *Leuciscus phoxinus.*
13. Tench, *Tinca vulgaris.*
14. Loach, *Nemacheilus barbatula.*
15. Eel, *Anguilla vulgaris.*
16. Sea-lamprey, *Petromyzon marinus.*
17. Planer's Lamprey, *Petromyzon branchialis.*

Next to the Lea, the Colne and Stour are the homes of more species of Fish than any Essex stream. At Sudbury, the fishing is well cared for and protected. The same may be said of Bures, Nayland, and Dedham. I think there are more followers of the gentle craft plying their sport on the Stour than anywhere else in the county, always excepting the Lea. This river has another charm for anglers, inasmuch as many of its coarse fish are plentiful and run to a large size.

I do not think Trout occur in it naturally—in fact, I never heard of the capture of a fish of this species in it at all, until the last few years. Again, excepting the Lea, wherever Trout occur in any of our rivers, they have been late introductions. Where introduced, however, they appear to have thriven well. Another species introduced in this river is the Wels (*Silurus glanis*), which does not appear to occur in any other Essex stream (cf. *Essex Nat.* vol. viii., p. 152). Probably, anglers generally will not be disposed to agree with the remarks of Dr. Günther, as quoted under its name in this list (see p. 106).

The following list may, I think, be relied on:

FISHES OF THE RIVER STOUR.

1. Perch, *Perca fluviatilis.*
2. Miller's Thumb, *Cottus gobio.*
3. Stickleback, *Gasterosteus aculeatus.*
4. Ten-spined Stickleback, *Gasterosteus pungitius.*
5. Trout, *Salmo fario.*
6. Pike, *Esox lucius.*
7. Carp, *Cyprinus carpio.*
8. Gudgeon, *Gobio fluviatilis.*
9. Roach, *Leuciscus rutilis.*
10. Dace, *Leuciscus vulgaris.*
11. Rudd, *Leuciscus erythrophthalmus.*
12. Minnow, *Leuciscus phoxinus.*
13. Tench, *Tinca vulgaris.*
14. Bream, *Abramis brama.*
15. Bream-flat, *Abramis blicca.*
16. Loach, *Nemacheilus barbatula.*
17. Eel, *Anguilla vulgaris.*
18. Wels, *Silurus glanis.*

The River Cam rises in the north-west corner of the county, but it soon leaves our district. Unfortunately, I have been quite unable to get any list or specimens of the fish of the Essex portion of this river, This is the more to be regretted, as, in some part of its course, it holds two species, apparently naturally absent from all the rest of our Essex rivers; and, for aught I know to the contrary, they ought perhaps to appear as members of our Fauna. These two species are the Grayling, lately introduced into the Lea, and the Spined Loach.

INTRODUCTION.

As regards this list of the Fish of Essex, I am fully aware how imperfect my efforts have been to provide a complete Catalogue. I have endeavoured to make it as reliable as possible, with the hope that its publication may be the means of directing the attention of other observers to our Sea and River Fish and, by so doing, may enable others, at some future period, to fill up my many blanks. It is quite impossible for anyone not residing at a fishing-port and with many other calls on his time to make, single-handed, anything like a complete record of all the species occurring on the coasts of a county so rich in species and individuals as Essex.

SUMMARY.

In conclusion, I may briefly point out that it is amongst the Seals, Cetaceans, and Fishes that we are most likely to have additions made to the lists comprised in this volume.

There is no reason why stragglers of at least two other species of Seals should not be recorded, as both are fairly common on the Norwegian coast. One of them, the Bearded Seal (*Phoca barbata*), has been added to the Norfolk Fauna during the last year: the other, the Ringed Seal (*Phoca fœtida*), may have been already frequently captured, and mistaken for the Common Seal. Turning to Cetaceans, it can hardly be possible that the Common Dolphin never visits our coasts; nor is it at all unlikely that some of the Rorquals or other Whales, absent up to the present time from our records, may be cast ashore here at any time.

It is in the Class Pisces that the greatest additions, however, will probably be made in the future to the Essex Fauna; for it is quite impossible that every species occurring on such an extensive coast as ours can have fallen under the notice of one or two observers; and, in this department of science, as I have already remarked, more than these few observers are not forthcoming in the county.

For convenience of comparison with those counties on the East Coast of England for which similar lists of Fauna have been published, the following table is appended. The last column shows the numbers for each of the corresponding classes, generally recognised as belonging to the British Fauna.

	Essex.	Norfolk.	Yorkshire.	Northumberland & Durham.	British Isles.
MAMMALS:					
Terrestrial	38	29	32	30-4 (?)	45
Marine	10	14	14	11	27
REPTILES:					
Terrestrial	4	4	4	...	9
Marine	...	...	2	...	2
AMPHIBIANS	6	6	6	...	7
FISHES	113	143	148	...	234
Totals...	171	196	206	...	324

LIST OF SPECIES.

Class MAMMALIA.

Order CHIROPTERA.

Sub-Order MICROCHIROPTERA, *Dobson.*

Family RHINOLOPHIDÆ, *Gray.*

Sub-Family RHINOLOPHINÆ, *Dobson.*

[*Genus* RHINOLOPHUS, *Geoffroy.*

Rhinolophus ferrum-equinum, *Schreb.* GREATER HORSE-SHOE BAT.

This bat is stated in both editions of Bell's *British Quadrupeds* and also in Cassell's *Natural History*, to occur at Colchester. I believe, however, that some mistake was made with regard to the locality of the specimens. The Horse-shoe Bats are so distinct in flight, and so unmistakable in the hand, that I do not think I am in error in affirming that neither occurs in this district. This is the more remarkable when we consider their abundance at Canterbury, but a short distance south of Essex. They are also plentiful in the western counties of England, and any number may be found in the caverns in the Mendip Hills. It must not be supposed that their absence from Essex is a question of climate, since they occur much farther north. The smaller species, *R. hipposideros*, has been taken in the neighbourhood of Ripon, in Yorkshire, which is probably its northern limit.]

Family VESPERTILIONIDÆ, *Dobson.*

Group I. PLECOTI, *Dobson.*

Genus SYNOTUS, *Keys and Blas.*

Synotus barbastellus, *Keys and Blas.* BARBASTELLE.

Now that I have learned to recognise it, I do not consider the Barbastelle so rare as it is usually believed to be. Still, I cannot call it a common bat. Doubleday says (*Zoologist*, 1843, p. 6) it is not uncommon in Epping Forest. I have found it, early in April, flitting slowly, and in an apparently purposeless manner, near the ground, under the protection of a plantation. This peculiar style of flight is one means by which it may be distinguished. It is very solitary in its habits, and haunts trees principally. I have always seen it away from the town; but, although it may have been noticed flitting under the cover of a hedge or plantation one night, it does not follow that it will be there if looked for on the following evening.

The ears of the next species are remarkable for their size: those of this species for the manner in which they are united across the forehead of the animal, a most unmistakable character, occurring in no other European bat. Its fur is also darker than that of any of our native bats, a circumstance which makes it appear larger than it really is. This sombre hue, and the peculiarity of the ears, will enable it to be readily recognised.

Genus PLECOTUS, *Geoffroy.*

Plecotus auritus, *Geoffroy.* LONG-EARED BAT.

This bat is very common in the Colchester district, and usually has its haunts in buildings, although it has been brought to me from hollow trees. I consider it equally common throughout all parts of the county. Doubleday says (*Zoologist*, 1843, p. 6) that it occurs at Epping. A mild and gentle creature, it is by no means difficult to tame. This is the bat most frequently caught in houses, which it enters by the open window, often much to the consternation of the female members of the household.

There is no difficulty in distinguishing this bat. Its remarkable ears, which are fully as long as its body, cause it to be unmistakable. No other existing animal, so far as I know, has ears in this undue proportion, except *Plecotus homochrous*, which occurs in the Himalayas and is a questionable species (cf. Dobson's *Asiatic Chiroptera*, 1876, p. 85). If this is only a variety, then the animal under consideration is, in respect of ears, unique. Although so large, the ears do not strike one as so disproportionate as those of a lop-eared rabbit, which in comparison are really much smaller.

Group II. VESPERTILIONES, Dobson.

Genus VESPERUGO, *Keys and Blas.*

Vesperugo serotinus, *Gmel.* SEROTINE.

This must be a very rare bat in Essex, as I have never had the good luck to meet with a specimen. Mr. Miller Christy, however, has been much more fortunate, and he has been able to record the occurrence of the only two examples ever noted in the county. The first he mentions (*Zoologist*, 1883, p. 173, and *Proc. Essex Field Club*, vol. iv., p. iv.) was in the possession of Mrs. Joseph Smith, of Great Saling, and had been shot, more than twenty years previously, in the garden of Pattiswick Hall, Coggeshall. The second (*Zoologist*, 1894, p. 423, and *Essex Nat.*, vol. viii., p. 162) he captured himself, it having entered his bedroom window at Pryors, Broomfield, on the night of the 25th August, 1894. These two captures appear to be the most northerly ones recorded for the species. Mr. Christy has taken every care that there should be no mistake in the identification of the specimens. The latter specimen, a male, he found to measure, in expanse of wing, fully fourteen inches. The earlier one, being badly stuffed and much shrunken, he was unable to measure.

Vesperugo noctula, *Keys and Blas.* NOCTULE OR GREAT BAT.

This occurs commonly at Colchester, as I believe it does throughout the county. It is also one of those noted by Doubleday (*Zool.*, 1843, p. 6) as occurring around Epping.

Bell says (*Brit. Quad.*, ed. 1874, p. 19) that it has a shorter period of activity than most of the order ; but how this misconception arose I cannot understand, since it may be seen on the wing from March to November. The latest period at which I have obtained a specimen is the 10th of November, but I have observed it even later than this during a favourable season.

This bat, the largest of the British Chiroptera, is, in colouring, the most beautiful of the order. Its rich brown fur, smoother and finer than velvet, contrasts strikingly with its black wings. Although it is rare for bats to vary much in colour, I once took a Noctule that was almost black. Like the rest of the family, it is of a quarrelsome disposition ; and, when irritated, its aspect betokens ferocity and savagery in a high degree.

The Great Bat well merits Gilbert White's name *altivolans*, as its flight is rapid and high, like that of the Swift. All through October, it may be noticed flying over Colchester and throughout the valley of the Colne, sometimes at a great height : at other times, it flies much below the tops of the houses in the streets.

Although trees are preferred by it as *hybernacula*, I know a few haunts where it may generally be found resting between chimneys and walls of houses. I never remember to have found another species hybernating with it.

Vesperugo pipistrellus, *Keys and Blas.* COMMON BAT.

This bat, a small imitation of the Noctule, is extremely common in all parts of the kingdom, and is probably the commonest of the order. It is found, of course, at Epping, and appears in Mr. Doubleday's list (*Zool.*, 1843, p. 6).

It is on the wing, in mild seasons, almost up to Christmas, when it finally retires for its winter sleep, choosing, almost invariably, for its home holes in and about buildings. Its flight, unlike that of its larger relative, is low, and well suggested by its vulgar name, "Flitter-mouse." The sheltered sides of buildings, or hedges, are its favourite hunting-grounds—no doubt in consequence of its food (gnats and small insects) seeking a like protection from the wind.

It sallies forth earlier and retires later than any other bat, and is the species most frequently seen flying by day.

Genus VESPERTILIO, *Keys and Blas.*

Vespertilio daubentonii, *Leisler.* DAUBENTON'S BAT.

Mr. E. Newman (*The Field*, Mar. 14th, 1874, p. 263), quoting from a letter of Doubleday's, says that he has seen this bat caught between Epping and Abridge, and has noticed it flying over the Stour at Sudbury, close to the water. In the Colchester district, it is not rare. In winter, I have found it, as did Yarrell, under the Castle at Colchester. Its usual habitat, when at rest, is in buildings.

It has one character by which it may easily be distinguished: *viz.*, the wing membrane, which extends only to the distal extremity of the tibia, leaving the foot free. In the other species, the membrane is usually continued to the root of the toes. Daubenton's Bat has been well described as an aquatic species. Its habit of haunting water, whether stagnant ponds or running streams, flitting with rapid vibrations of the wings a very few inches above the surface, and apparently spending most of its time of flight there, makes the description "aquatic" a very exact one.

Those I have endeavoured to keep appeared delicate, and soon died. Possibly they required aquatic insects, or it may be they could not bear confinement. The members of the entire family appear equally unfitted for close captivity, and I have never succeeded in keeping insectivorous bats alive for any length of time.

Vespertilio nattereri, *Kuhl.* REDDISH-GREY BAT.

This easily-distinguished bat is reputed to be rare. In the district round Colchester, however, it is one of our commonest species. It is recorded by Doubleday (*Zool.*, 1843, p. 6) as found at Epping, and Newman also notes (*Field*, Mar. 14th, 1874, p. 263) that it is not uncommon there. I hear the peculiar squeaking note of this bat during its evening flight more frequently than any other in the streets and gardens of Colchester.

Houses and buildings are its favourite hiding-places in summer. In winter, cellars, caverns (as those under Colchester Castle), and similar places are generally chosen. Occasionally, in the late autumn, bats are brought to me which have been drawn up to the surface in buckets from some of our deepest wells. From my observation that three out of every four are of this species, I am led to believe that crevices in the brickwork of the wells are chosen by them as *hybernacula*.

Vespertilio mystacinus, *Leisler.* WHISKERED BAT.

The flight of this bat is very similar to that of the Pipistrelle; and like that animal, it hawks under the shelter of a hedge or row of trees, where I have no doubt it is frequently passed over as the commoner species by those in search of it.

Like the last, it is much more common than is generally believed. I have had no difficulty in finding all I have required for the purposes of study. Doubleday says (*Zool.*, 1843, p. 6) it is found at Epping, and I have no doubt it is distributed throughout the county, for I have seen it wherever I have looked for it.

Order INSECTIVORA.

Family ERINACEIDÆ, *Bonap.*

Genus ERINACEUS, *Linn.*

Erinaceus europæus, *Linn.* HEDGEHOG.

I should think there are very few places in Essex where this very common animal is not to be found. It may be seen in woods, hedges, and in the coarse herbage about the ditches of the marshes.

The persecution it has long undergone at the hands of the game preserver, while diminishing its numbers, happily has not caused its extermination. Doubtless it sometimes indulges in the theft of an egg, a young partridge, or a pheasant; yet these delinquencies must be overlooked in consideration of the hedgehog's extreme usefulness in destroying grubs and insects. Numerous instances are also recorded of its having undoubtedly made a raid upon the poultry, and carried off young chickens from the hen; while gamekeepers very generally assert that it carries on a similar practice with young pheasants. Even so, I think we must still consider it to be one of the most harmless animals we have.

In confinement, the hedgehog is quick to lose all fear, neither curling up, nor erecting its spines when handled, and soon learning to come to be fed. This is especially the case, if the creature is kept in the kitchen, where it may suitably occupy itself in destroying the cockroaches which sometimes frequent that apartment.

Family TALPIDÆ *(Gray)*. *Dobson.*

Genus TALPA, *Linn.*

Talpa europæa, *Linn.* COMMON MOLE.

This useful animal is common in all parts of the county, although in every district it is subjected to great persecution, since the farmers naturally object to its practice of burrowing and casting up mounds in their pastures and newly-sown fields. I am not aware of any other mischief done by this creature in Essex, except, it may be, that it disarranges and blocks up the land-ditches by burrowing under them. In other counties, no doubt, considerable damage has been done by this industrious little miner, in boring through embankments made to keep out water, thus necessitating every means to be taken to keep down its numbers.

The hillock which covers the nest of the mole is generally made in a hedge or wood, although it lies sometimes quite in the open field. It may generally be known by its size, being much larger than the ordinary molehill. The young are from three to six in number, and are produced in the spring. They are born naked, but grow rapidly, and are soon covered with the fine fur or down. While young, they are able to fast a considerable time. I have had some brought to me alive which had been taken from the nest on the previous day, and had remained without food for many hours, a privation which would speedily have proved fatal had they been full-grown.

The cream-coloured variety of mole is not rare, especially on some farms.

There is, near Colchester, a family of mole-catchers, the greatest masters of their art whom I have ever known. They will, at any time, produce living specimens at a few hours' notice. They bear the appropriate surname of "Watchem," originally no doubt, a nick-name given them in consequence of their trade. Of late years, the family have altered the spelling to Watsham. I have purchased of one mole-catcher (who works in company with his brother) as many as fifteen hundred fresh skins in a season, from which fact the enormous scale on which moles are destroyed by an able man may be surmised.

Family SORICIDÆ, *Linn.*

Genus SOREX (*Linn.*), *Wagl.*

Sorex vulgaris, *Linn.* COMMON SHREW.

This animal occurs commonly in all parts of the county, although it is more frequently heard than seen. Like all the rest of the family, it is highly pugnacious, and two rarely meet without engaging in a fight. Hedge and coppice, in consequence, frequently resound with their faint but shrill war-shrieks. In colour, it varies very much, two specimens hardly ever occurring of exactly the same shade.

Sorex minutus, *Linn.* LESSER SHREW.

I find *Sorex vulgaris* and *Sorex minutus* equally common, either as captured specimens or lying dead on the paths in autumn.

There can be little doubt of the specific distinctness of this from the last. The most obvious character by which to distinguish them is the comparative length of tail. The colour of the tips of the teeth is not so important a point, as it varies much with age. The hair on the lower portion of *Sorex minutus* will, I think, generally be found to be of a clearer white than is the case with the other species.

Genus CROSSOPUS. *Wagl.*

Crossopus fodiens, *Wagl.* WATER SHREW.

This interesting little animal occurs in all parts of the county, in suitable localities. It usually prefers shallow stagnant pools to quick-running streams, but is occasionally found at some distance from water. There is a small pond on Stanway Heath, at least half a mile from any other water, where I often see these animals.

This Shrew is an expert swimmer, but seems to prefer running about underneath the water. This it does as freely and with as great apparent comfort as on dry land, using, meanwhile, its long snout to turn over any substance which may hide its prey. Its food consists chiefly of beetles, their larvæ, and other insects, as well as fresh-water crustaceans.

Sorex remifer (the Oared Shrew), formerly considered distinct, is, according to the latest authorities, only a dark form of *Crossopus fodiens.*

Order CARNIVORA.

Family MUSTELIDÆ.

Genus MELES, *Stow.*

Meles taxus, *Boddaert.* BADGER.

This animal, known so well to most of us by hearsay from a previous generation, has now become scarce in the county, where, fifty years ago, it was very common. Clearance of woods, diminution of hedgerows, and excessive game preserving have been the most effectual causes of its decrease. The latter, as it is now practised, will, I fear, in time blot out not only the poor Badger, but every other animal which, whether useful or not, can be classed by ignorant gamekeepers or their masters under the head of "vermin."

Mr. E. A. Fitch has published (*Essex Nat.*, vol. i., p. 186) some extracts from the churchwardens' accounts of the parishes of St. Peter's and All Saints, Maldon, showing the Badger to have been very common in that borough (which then included some woodland) from the year 1716 to 1754, and probably long afterwards. The reward paid for the destruction of this "vermin" appears to have been uniformly a shilling a head, the same sum as was paid for a Fox.

The quantity of bones of badgers found in exploring the Dene-holes at Grays (*Essex Nat.*, vol. i., p. 257) points to its having formerly been a common animal in that district.

Within the last few years, two Badgers have been taken in the Colchester district; and, still more recently, another was caught at Bentley, between Colchester and Harwich. One almost feared that these might be the last survivors of a race which gave much sport to our ancestors—or, perhaps, one should say, gave opportunity for the exhibition of much brutality. In my younger days, I saw a few baitings of Badgers caught at South Benfleet, where, in 1844, there were several earths. It did not strike me, however, that any great amount of suffering was inflicted on the Badger. The dogs, especially those new to the work, gave unmistakable evidence of their dread of the Badger's jaws. Mr. J. E. Harting quotes (*Essex Nat.*, vol. iii., p. 197) a song from *The Sportsman's Vocal Cabinet* (1830, p. 136), edited by Charles Armiger, describing the delights of a Badger hunt in Epping Forest. The unfortunate beast was brutally ill-treated when captured.

Still, the Badger is by no means extinct in Essex. The fortunate preservation of Epping Forest by the Corporation of London has provided a safe retreat for some pairs which Mr. E. N. Buxton introduced into the woods in 1886. Mr. Buxton says (*Epping Forest*, 4th Ed., 1897) that they now occupy a large "holt" in Loughton Manor, and are increasing, extending the earth every year. " They draw out astonishing mounds of soil from the holes, and their underground township must be very roomy. They are tenants in common with the foxes. Indeed it is an old fox-earth. Rabbits also occupy a portion." Mr. Buxton adds that, although they generally live in harmony, there is evidence that the Badgers "are occasionally guilty of the heinous crime of vulpicide." The accompanying sketch by Mr. H. A. Cole represents the "earth" in which the Badgers here mentioned live.

Some of the following records may be of stragglers from the Forest District:

Mr. James English says (*Essex Nat.*, vol. i., p. 183) that one was killed at Hill Hall Wood, Theydon Mount, in 1850; another was shot in the same wood in 1874. Mr. Fleming records (*Field*, March 14th, 1874) the capture of a fine animal at Shailsmoor Spring, in this county, within seventeen miles of London. Twelve years previously, near the same place, one was dug out of a rabbit burrow. The next occurrence shows that the species is not quite so invulnerable as we could wish, since Mr. Miller Christy informs me that Mr. G. H. Baxter, of Hutton Park, has a stuffed female badger which a terrier killed in Epping Forest, near the Rising Sun, Woodford, about the year 1888. The same gentleman records (*Zool.*, 1882, p. 303) one found dead in New Pasture Wood, Great Saling; and another discovered dying in High Wood, Writtle Park (*Proc. Essex Field Club*, vol. iv., p. lxviii). The latter's existence had long previously been known to the captor. Mr. E. A. Fitch notes (*Essex Nat.*, vol. i., p. 114) that a pair took up their abode in Park Wood, Shalford, and bred there for at least three years. Mr. Robert Lodge mentions (*Essex Nat.*, vol. i., p. 186) one shot by mistake for a cat in an outhouse on the north side of Ilford Station in August, 1887. Another was killed on the line near Theydon

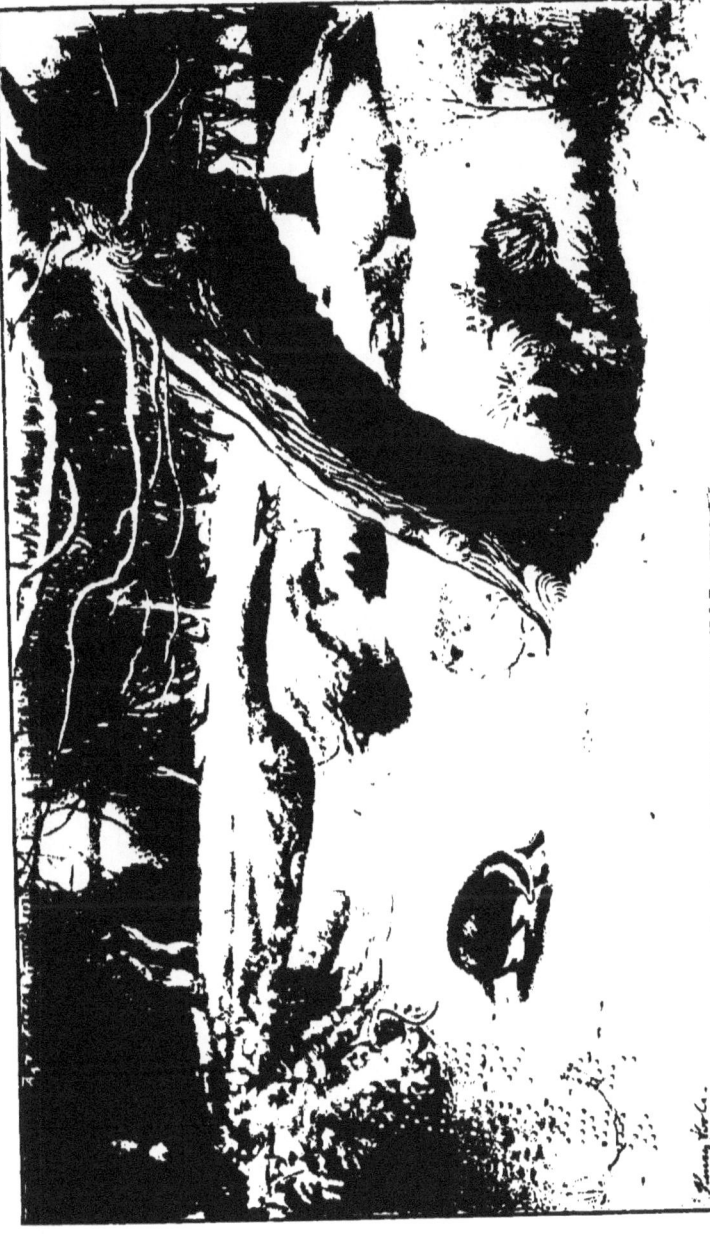

Badger Earth In Epping Forest.

Bois station by a train (*Essex County Chronicle*, November 9th, 1888). Several more were reported from the Epping Forest District at the same time.

In Mr. Philip Benton's *History of Rochford Hundred* it is stated (vol. i., p. 197), that, so long ago as 1841, one was captured on Foulness Island, and another on New England Island. Both had burrows in the sea-wall. In May 1891, a Badger was unearthed by a rabbit-shooting party on the Asheldham Hall estate (*Essex County Chronicle*, May 29th).

Mr. B. Morris, writing in 1894 to a London newspaper (*Standard*, May 17th), related that eleven Badgers had been captured in a wood near Braintree shortly before, and expressed his regret that there should have been such a wholesale destruction of this interesting animal. It appeared, however, from further correspondence published, that only two of the number were killed, the remaining nine being set at liberty in various parts of the kingdom, some being retained in the county.

In 1896, I purchased a pair of young Badgers, lately caught with their mother (who was unfortunately killed), at Stanway, where Badgers appear to have continually existed. They were liberated in the county on an estate where they will be well cared for. One, weighing 28 lbs., was shot in Brooke's Wood, Stisted, early in May, 1897 (*Essex County Chron.*, May 14th, 1897).

Remembering the above records, we may hope that the Badger will long continue to be a member of the Essex Fauna.

Genus LUTRA, *Erxleben*.

Lutra vulgaris, *Erxleben*. COMMON OTTER

This animal is not as uncommon in Essex as it was in years gone by, when there seemed every probability that the Otter would become quite extinct in the county. Its disappearance may be accounted for, not by the increased value of the coarse fish upon which it subsists, so much as the comparative ease with which it may be taken in a steel trap, and ultimately converted into a "specimen" to adorn the hall wall—a horrible "stuffed" effigy of its former graceful self. One cannot sufficiently regret the constantly-recurring notices in the public prints of Otters having been shot. Yet these notices prove that the animal is now to be found all over the county, and the very frequency of the announcements shows it to be present in greater numbers than at one time. It is possible that it is not now diminishing in number, for the capture of Otters is reported not only from the rivers but also from the reeds and sedges of the marshes, where they had not been seen for years. In the Stour, Chelmer, Blackwater, and Lea, Otters occur frequently.

The Rev. W. B. Daniel in his *Rural Sports* (1812, vol. i., p. 620) says that a gigantic Otter, weighing upwards of forty pounds, was snared in October, 1794, in the River Lea, between Ware and Hertford. He also mentions (*op. cit.*, p. 631) nine killed in one day in the frozen "fleets" of Fobbing Marshes. Mr. Arthur Fitch, of Wixoe, shot two females in the Stour on 24th December, 1888; and Mr. E. A. Fitch reports (*Essex Nat.*, vol. i., p. 105) quite a number from the neighbourhood of Maldon, both in the Chelmer and Blackwater rivers. Others at Rayne, Sturmer, and Baythorne End are also noted (*ibid*, 280).

Mr. French relates that a mother and two young Otters were taken alive in the Chelmer at Felstead, in February 1891. They were returned to the river (*ibid*, vol. v., p. 73).

In August 1891, a young female was caught alive in the Tendring Hall Brook, and was forwarded to the Zoological Gardens (*Essex County Chronicle*, August 14th, 1891). At the end of the following November (*ibid*, Dec. 4th), a male Otter, weighing 23 lbs., was found in a brook at Great Hallingbury by a fox-terrier, whose owner was compelled to shoot the Otter to save the dog, which it savagely attacked. A fortnight or so later (*ibid*, Dec. 24th), a pair were shot in a "fleet" close to the Cottage, Great Bentley.

In the *Essex Naturalist* for 1893 (vol. vii., p. 123) is a record by Mr. Bateman of Otters breeding in the open near Brightlingsea; and, in the same periodical (*Essex Naturalist*, vol. vi., p. 138), Mr. W. Cole relates how a baby Otter found in a rabbit's hole at East Mersea was suckled by a cat with her kitten.

Mr. J. Surridge, jun., writes (*Field*, 18th April, 1874, p. 374) an amusing account of an impromptu otter hunt in the park at Stisted Hall, the animal being apparently both strange and terrifying to the keepers and servants of that estate during the absence of the family.

Daniel also relates (*Rural Sports*, vol. i., p. 625) that Mr. Edwards, of Little Waltham Hall, owned an Otter which attended him like a dog, and which, every afternoon, while the old gentleman slept, regularly stationed itself upon his lap. It obtained fish from the various ponds in the gardens and grounds near the house, and was fed also upon milk. At last, it was accidentally killed by a maid-servant striking it with a broom-handle upon the nose, where the slightest blow is fatal.

The Otter is certainly one of our most interesting and graceful animals when seen swimming in its native streams. It is astonishing that a creature of its size should be able to slip in and out of the water so quietly, and without making half the "wake" that a Rat does. Anyone fortunate enough to see, as I have more than once seen, a mother Otter and her family playing in clear water, will, I am sure, agree with me that it is one of the most attractive aquatic sights possible.

Otters, like the rest of the family, are nocturnal in

their habits. I have heard from an elderly friend that on one occasion, when the ground was completely covered with snow, he tracked an otter for some miles, upon its passage from pond to pond. This distance it had travelled during the night. The district was Dengie Hundred, and the time that when ponds containing fish, or at all events eels, were much more abundant than now. As is well known, almost every field formerly had its clay-pit or pond, now drained and filled up in consequence of changes in agriculture.

A very interesting article* has been published by Mr. John Watson, on the food of the Otter. The author clearly shows that, as the dietary of this animal is extremely comprehensive, its depredations upon the stock of fish contained in our rivers cannot really be serious.

Genus MUSTELA, *Linn.*

Mustela vulgaris, *Erxleben.* COMMON WEASEL.

In spite of persecution, this small animal is common in all parts of Essex, and especially on the marshes, where I have found abundant evidence in casts that it not unfrequently forms the food of Herons.

The Weasel may be distinguished at a glance from the Stoat by the colour of its tail, which is of the same reddish-brown tint as the upper surface of its body. An additional distinction may be usually found in the smaller size of the Weasel. This character, however, is not immutable, as I have seen Weasels quite as large as an average Stoat, and full-grown Stoats as small as under-sized Weasels.

This is another of the so-called "vermin," and is ruthlessly destroyed whenever it is found. This, in my opinion, is a great mistake, since its prey consists chiefly of the smaller

---

* "Water Poachers," *Nineteenth Century*, Oct. 1889; *cf. Essex Nat.*, vol. iv., p. 84.

Mammals, which might otherwise become too numerous. Its food is not confined to these, for I have found Beetles, Lizards, Slow-worms, and other small fry in the stomachs of those examined. It is rare for the Weasel to interfere much with game.

In some parts of England, the Weasel, when small, is called a "mouse-hunter," and right well it deserves this name. When chasing a mouse, it keeps to the scent as well as the best foxhound, and seems regardless of onlookers. I have said "keeps to the scent"; but it would be more accurate to say that it never passes over the scent, even when in full gallop. Its method appears from observation to be that of making casts diagonally across the track of its prey. This plan of hunting, I have an idea, is not confined to the Weasel.

Although albino specimens of this species are rare, Mr. Miller Christy has, on the authority of Mr. J. Pettitt, taxidermist, of Colchester, recorded (*Zool.*, 1895, p. 19), one killed near that place about December 20th, 1892. It was pure white, had pink eyes, and its flesh was of an unusually pale colour.

Mustela erminea, *Linn.* STOAT.

This rapacious, active, and destructive member of a specially blood-thirsty family, is very common throughout Essex. Persecution, from every quarter, seems to have little influence in diminishing its numbers.

Its food and habits are similar to those of the Weasel, but I fear I cannot defend it from the charge of being very destructive to game. Although mice and small creatures contribute to its sustenance, its favourite prey consists of the larger mammals. Hares, Rabbits, and Rats are chased by it from scent, as most persons who live in the country can testify from observation. The two former, after being pursued for some distance, appear to resign themselves to their fate, without further effort to escape. Whether they become paralysed by fear or exhaustion, or by both, I am unable to state; but, after a faint struggle, they certainly sit still and allow the wily little hunter to attack, offering no other defence than screams.

The following note of a fight between a Weasel and a Stoat seems worth recording, as it is rare that a witness is present at such an unusual encounter. The quarrel probably originated in the appropriation by one of the combatants of the freshly-captured spoils of the other:

"On Saturday, as Mr. Kebby, of Beeches Farm, Stock, was walking round one of his fields, he heard a screaming noise, and on going to the spot he saw a large male Stoat and a male Weasel in the ditch fighting. The Stoat had the Weasel in its mouth trying to rend it. Mr. Kebby jumped into the ditch, and killed both combatants. The Stoat and Weasel have been carefully preserved in the attitude they were seen fighting in by Mr. C. Cable, naturalist, of Stock" (*Essex County Chronicle*, November 28th, 1890).

Stoats in winter livery have been often seen in the county. One such was shot by Mr. Brundish, at Willingale, in March, 1871, and I have seen several which were perfectly snow-white. More frequently, however, they retain some reddish patches on the head and shoulders.

The number of young varies from four to five, and more playful little creatures than a family of young Stoats it would be difficult to find.

Mustela putorius, *Linn.* POLECAT.

This animal, so destructive both to poultry and game, is becoming very rare in Essex. In many districts, it is quite extinct, even in spots where, only a few years since, it was frequently seen.

Mr. H. M. Wallis says (*Zool.*, 1879, p. 264): "My father remembers seeing five full-grown Polecats killed together in a drain, by a terrier, near Chelmsford."

Mr. Miller Christy writes: "Mr. William Raeburn informs me that he saw, in the possession of the host of the Tower Arms inn, at South Weald, a Polecat killed there about twenty years ago."

Mr. Reginald W. Christy reports (*Essex Nat.*, vol. ii., p. 37): "The last specimen known to have been killed in the neighbourhood of Roxwell was trapped on the Boyton Hall Farm in or about the year 1855."

The food of the Polecat is as varied as that of the other members of the family, and embraces, according to some authorities, fish, frogs, and other reptiles.

There is not much difference in appearance between a dark Ferret and a Polecat; and the probability is that the Ferret is simply a domesticated Polecat, but domesticated in a warmer climate than ours. This, no doubt, accounts for the greater susceptibility to cold of the domesticated animal. Part of this tenderness is due, doubtless, to the conditions under which Ferrets are reared. Mine, reared in an open pig-court, are never the shivering animals born in warmer situations. On the contrary, I have observed them to roll and tumble in snow, apparently without discomfort, if not with enjoyment.

Mustela martes, *Linn.* COMMON MARTEN.

The Rev. R. Lubbock, in his *Fauna of Norfolk* (1845), says this animal is still occasionally found in Essex, and there is good ground for hoping that the graceful and active creature yet exists in the county. It was formerly very common, and I have heard old sportsmen speak of shooting it from deserted magpies' nests.

Mr. H. M. Wallis says (*Zool.*, 1879, p. 264): "In 1822, one was killed in the Waltham Woods, near Chelmsford, by Mr. Thomas Gopsill, of Broomfield." Mr. Harting, writing in 1880, says (*Trans. Essex Field Club*, vol. i., p. 95), the last killed in Essex, so far as could then be ascertained, was trapped in April, 1853, by Mr. Luffman, head keeper to Mr. Maitland, in one of that gentleman's covers at Loughton. This specimen was examined by Mr. English, who afterwards recorded (*Journ. of Proc. Essex Field Club*, vol. iv., p. lxiv) having seen another in the forest near Ambresbury Banks, on July 20, 1883. From all accounts, it would appear that the Marten does still remain in the Forest, other evidence corroborating the accounts of Mr. English. Mr. E. A. Fitch records (*Essex Nat.*, vol. iv., p. 153) the existence of undoubted Martens in Hazeleigh Hall Wood, although he himself failed to capture an individual. The same gentleman reports another instance in Epping Forest, although about this last some mystery appeared to hang (*Essex Nat.*, vol. iv., p. 126). They are not found in any great numbers.

Daniel, in his *Rural Sports* (1812, vol. i., p. 608), says: "When taken young, the Marten is easily tamed, and is extremely playful and good-humoured. Its attachment, however, is not to be relied on if it gets loose, for it will immediately take advantage of its liberty and retire to the woods, its natural haunts. A farmer in the parish of Terling was famous for taming this animal, and had seldom less than two in his keeping. Some years since [from 1801], one used to run tame about the kitchen of the Bald-Faced Stag inn, in Epping Forest. . . . The most ever met with by the compiler was in the large woods near Rayleigh."

There were supposed, at one time, to be two species of British Martens; but Mr. Alston has since satisfactorily proved (*Zool.*, 1879, p. 441) that we have in this kingdom a

Laver's "Mammals, etc., of Essex." [*To face p.* 51.

Fox Earth in Epping Forest.

single species only of Marten, and that the one whose name appears above.

Family CANIDÆ.

Genus CANIS, *Linn.*

Canis vulpes, *Linn.* COMMON FOX.

It is quite unnecessary to give any specific account of this fortunately-common member of our Fauna, which is familiar to almost everyone.

I say "fortunately-common," and long may it remain so; for the sport of which it is the object does great good in bringing all classes together, encourages the breeding of horses, trains our young men as fearless riders, and does not make its votaries selfish and suspicious, as is the case with shooting, fishing, and most other sports. So far as I know, the pursuit of the Fox and its preservation are the causes of no serious damage to anything except to a little poultry and game. The advantages, on the contrary, are so manifest that we must be content to lose these in exchange. So long as there are woods in the country—and the present condition of agriculture gives no reason to suppose they will be destroyed—so long will there be Foxes, unless the game preserver takes to using poison, and so effectually destroys the sport of the many for the sake of a day or two's grand battue during the season for a few. This I hope we may never see.

A drawing of a Fox's Earth—one of the many to be found in Epping Forest—from the brush of Mr. H. A. Cole, faces this page.

The cunning of the Fox is proverbial. Daniel, in his *Rural Sports* (1812, vol. i., p. 273) relates one or two instances observed in this country of its extraordinary sagacity, and of its tenderness for its young. Thus, one was observed to drop a cub from its mouth, after it had been hotly pursued for many miles in the neighbourhood of Chelmsford.

Another instance was in April, 1784, when Daniel's own whipper-in, returning home after being thrown out by accident, was induced, by the action of a terrier which accompanied him, to examine more closely a pollard oak tree near a cover on Broomfield Hall farm. Climbing the tree, he discovered a Fox and four cubs in a deep hole, at least twenty-three feet above the ground. She had apparently littered there, and had no other means of reaching them than by climbing the stem of the tree, which was thickly covered with twigs. Many people, the author goes on to say, inspected the tree, and three of the cubs were reared up tame to commemorate the incident.

Of the Fox's cunning, Daniel relates (*Rural Sports*, vol. i., pp. 257-258) a humorous story: In 1785, Mr. (afterwards Sir Henry) Bate Dudley, who hunted the Dengie Hundred country with his hounds, frequently had "a good Drag" on the banks of the river Crouch without finding a Fox. One morning, as they were drawing the remote churchyard of Cricksea, strangely overgrown with thick blackthorn bushes, a labouring man called out to the huntsman, "You are too late to find Reynolds at home. He crept off when he heard the hounds challenge about a quarter of an hour ago." In consequence of this information, the hounds were taken to different spots for some miles around, but a fall of sleet prevented their finding that day. A fortnight after, however, the Fox was found in an adjoining copse, and after a smart run of over two hours, he shaped his course to his favourite churchyard, where, apparently, desolation and neglect reigned supreme. The hounds being there at check, one of the pack suddenly reared itself up against an ancient buttress and gave tongue; whereupon the master, declaring his reliance on this, a favourite hound, dismounted and climbed up the broken buttress to the low roof of the church. Here, among the wild overgrowth of ivy, they found five or six fresh "kennels." Every one present quickly entered into the "spirit of the hounds," and promptly hauled them up the buttress, when three or four couple were in an instant in full cry on the chancel roof. "There," adds Daniel, "this extraordinary Fox was compelled to surrender up his life without benefit of clergy"—an event which he commemorates by printing a poem written by a Mr. William Pearce.

In that popular book *The Essex Hunt*, by Messrs. R. F. Ball and Tresham Gilbey (London, 1896, 4to), many notable runs with hounds in Essex are described. One or two of these may be briefly alluded to.

On February 19th, 1863, shortly before the resignation of the veteran master, Mr. Joseph Arkwright, a late-found Fox from Curtis Mill Green, near Navestock, ran through Pyrgo, the Havering Woods, over Upminster Common, and towards Dagenham. Before reaching this place, however, he turned and made for Hainault Forest, and again changing his mind, he dodged back, and headed towards the Bower Wood, at Havering, where he was killed after a splendid run of three hours and seven minutes, covering a distance of not less than twenty-four miles (*op. cit.*, p. 141). A run of quite a different character is described by Mr. Ball (*op. cit.*, p. 174). A Fox found in Parndon Wood, near Harlow, on Saturday, March 12th, 1881, ran towards Nettleswell and Cheshunt, then back through Galley Hills and Deer Park, on to Nazing Common, and so, returning to Parndon Wood, he went to ground in his old lair. He was dug out and killed after a fast run of one hour and forty-five minutes, with no check of more than three minutes. The find being very quick, only seven got away. Another wonderful run of three hours occurred on November 24th, 1890, from Kelvedon Hall Wood, the Fox being killed in Brentwood High Street.

54 THE MAMMALS, REPTILES, AND FISHES OF ESSEX.

Lord Rookwood supplies to the same volume (pp. 172-174) some interesting reminiscences of Foxes in strange situations. In the season of 1879-80, more than one was discovered, he recounts, in a certain ivy-clad tree in his own gardens at Down Hall. The animal was once quite forty feet above the ground, and on another occasion three were seen at the same time on branches of the same tree, from whence they were dislodged by aid of a ladder and a long pole. During the following season,

SUPPOSED "WOLF" FROM ONGAR WOODS.

the beast's cunning suggested to him to make a bolt up the chimney of some outbuildings at Fyfield, where his ears, and finally his head, were seen protruding from the top, cautiously observing the men and horses in the street. He then sprang out on the other side of the roof, with the pack behind him, but got away safely after all, and was lost.

[Canis (*sp. incert.*). JACKAL.

A specimen of a supposed "Wolf" from Ongar Woods (where it was taken after 1862) is now in the Essex Field

Club's Museum at Chingford. It was reputed to have been imported with young fox cubs. A notice of it, by the late Mr. Joseph Clarke, F.S.A., is printed in *Journ. Proc. Essex Field Club* (vol. iv., p. ccviii). A sketch of this specimen, by Mr. H. A. Cole, is here produced. Another notice of the supposed occurrence of this animal in Epping Forest appeared in *Land and Water* (July 19th, 1884, p. 64). It transpired afterwards that the experts had made a mistake; for further examination of evidence proved that the so-called Wolf was, in reality, a North African Jackal (*Journ. of Proc. Essex Field Club*, vol. iv., p. cciv.). Whatever the animal may have been, it cannot be claimed as a legitimate member of our Fauna; but, as there are very probably other individuals of the same species in existence in the Forest, it can hardly be passed over without mention.]

Sub-Order CARNIVORA PINNIPEDIA.
Family PHOCIDÆ.

Genus, PHOCA, *Linn.*

Phoca vitulina, *Linn.* COMMON SEAL.

This Seal occurs sparingly on all parts of the Essex coast, but it is not seen every year. Specimens have been killed in the Stour, the Blackwater, in the mouth of the Thames, and in other places. Properly speaking, all the Seals taken on our shores can only be considered as stragglers.

Two were observed in the Stour, between Harwich and Manningtree, in 1854 (*Field*, March 11th, p. 220). The young one was shot by a Stour puntsman near Mistley; but the older Seal, after being wounded by Mr. F. G. Folkard, was abandoned, and probably drifted on shore.

Mr. E. A. Fitch records (*Essex Nat.*, vol. ii., p. 3) the capture of two in the Blackwater and also of one in the Roach river. The former were shot by a Maldon fisherman on January 19th, 1881, the day following the remarkable snow-storm, when, as Mr. Fitch points out, the condition of the river Blackwater, choked with snow and ice, strangely resembled the native haunts of the Seal. That taken in the Roach was shot by Mr. J. G. Wiseman, of the Chase, Paglesham, on October 17th, 1887. It was said to have been forty years since a Seal had been previously shot in that river.

One was shot in the Thames, off Leigh, in the beginning of November, 1894 (*Essex County Chronicle*, November 9th, 1894).

About June 16th, 1894, a Seal was shot by Mr. William Linnett, off St. Peter's Chapel, Bradwell-juxta-Mare. Messrs. E. A. Fitch and Miller Christy, who saw the skin shortly after in the possession of Mr. James Spitty, of the Waterside, Bradwell, were informed that its length, when freshly killed, was 4ft. 6in., and its weight 80lbs.

Genus CYSTOPHORA, *Nilss.*

Cystophora cristata, *Erxleben.* HOODED SEAL.

Mr. W. B. Clark records (*Zool.*, 1847, p. 1870) the capture of a specimen of this Seal, on June 29th, 1847, in the Orwell. He gives a full description of the animal, which was presented by Mr. Ransome to the Ipswich Museum, where it now remains. As the river Orwell empties itself into Harwich Harbour, I think we are entitled to place this Seal in our catalogue of the Essex Fauna, although its normal habitation is within the Arctic circle.

Genus HALICHÆRUS, *Nilsson.*

Halichærus gryphus, *Fabricius.* GREY SEAL.

Mr. Southwell has drawn my attention to a record of the capture of this Seal in the Colne, printed in the *Annals and Magazine of Nat. History* for 1841. The incident had occurred a few years previous to that date.

By permission of the Editor, I published in the *Zoologist* (1885, p. 108), a record of this capture, which was made, I remember, by some fishermen in their nets. The specimen, which was aged and blind, was dissected by Professor Clark, and presented to the Cambridge Anatomical Museum.

A great source of error in consulting records of Seal captures is the uncertainty about the species, in consequence of the finds being rarely examined by competent naturalists. With more observers, too, no doubt some of the other species of Seals would have been added to this list.

Order RODENTIA.

Family SCIURIDÆ.

Genus SCIURUS, *Linn.*

Sciurus vulgaris, *Linn.* COMMON SQUIRREL.

This elegant and active little animal is so well known that very little need be said about it. It occurs in all parts of the county where suitable spots (that is woods) are to be found.

It is almost omnivorous. Mr. E. A. Fitch gives (*Essex Nat.*, vol. ii., p. 71) some proofs that it occasionally kills small birds for food. Birds' eggs and insects are also sometimes eaten, but vegetables form its main support. In the early spring, when the beech-trees are coming into leaf, I have noticed as many as six squirrels busily feeding on the young shoots of trees. They will bite off and throw down the leaves, consuming only the stalk—that is, the young branch. In the autumn, I have seen Squirrels strip off the loose bark from dead branches, and carefully scrape out with their teeth the fungus frequently found under the bark. When this ingenious process is going on, the position adopted by the Squirrel is not the usual one for feeding; for instead of sitting up on its haunches, the creature almost invariably hangs head downwards.

Although said to hybernate, I question if the Squirrel, as

a rule, does so. It probably takes a prolonged sleep; but, be the weather ever so cold, I have always found a Squirrel on the move, when it is properly looked for.

Family MYOXIDÆ.

Genus MYOXUS, *Schreber.*

Myoxus avellanarius, *Linn.* DORMOUSE.

The Dormouse occurs in those parts of Essex where the oak and hazel abound, and where there is sufficient woodland or overgrown hedgerow to protect it. It was formerly very common on the roadside at Berechurch; but, after the severe winter of 1860, the numbers were greatly diminished. During that winter, I found many nests in the bushes containing dead occupants.

South of the river Stour (that is, on the Essex side), Dormice are everywhere to be found on the higher ground above the meadows; but, on the north or Suffolk side, they appear to be unknown from Shotley until Long Melford is reached, where again they are not rare.

Mr. Rope gives (*Zool.*, 1885, p. 201) the range of this species in Britain. Its distribution seems as unaccountable as that of the nightingale.

This creature is one of the best examples of a hybernating quadruped in the country. As a rule, I think the winter sleep is taken underground, the bush nest not being used for that purpose. This may frequently be found deserted. If the mouse is disturbed in its bush nest, it is extraordinarily quick in its movements among the twigs, thus forming a strong contrast to the apparently sluggish creature usually seen in confinement. Its food is very similar to that of the Squirrel, but I have occasionally taken them licking up the "sugar" I had placed on tree trunks for the purpose of attracting moths at night.

Family MURIDÆ.

Genus MUS, *Linn.*

Mus minutus, *Pallas.* HARVEST MOUSE.

This very beautiful and active little creature occurs in all parts of Essex. In the winter time, it is found in corn-stacks, especially those placed in the fields, and most frequently, I think, in oat-ricks. I never discovered more than a dozen in one rick, although others have informed me that they have been found more abundantly in such situations.

As a pet, the Harvest Mouse is very interesting, and rarely quiet, day or night. They are very peaceable all through the winter, and any number may be kept together; but, in the spring fighting goes on until all, or nearly all, the males are destroyed and eaten, for they are dreadful cannibals. On the whole, I can strongly recommend them as pets. They are sweet, not at all mousy in odour, and very amusing in their ways. The longest time I have had them in confinement is over two years, but I never could make them so tame as my pets of the next species (*Mus sylvaticus*). I have taken every precaution possible, but have never been able to get them to rear their young in confinement. All might proceed well for a few days, and then their cannibal tastes would be indulged in, and the entire brood of young would be destroyed and eaten.

Indoors, Harvest Mice do not become torpid; nor do they when living in corn-ricks. I have never found any young in corn-ricks, although they are said to breed there. I consider their breeding-season is entirely confined to the summer months. This habit perhaps helps to prevent them becoming the pests to the farmer and gardener that some of the other mice undoubtedly are. Their numbers also are so small, they never can do much damage; and, as their diet is composed largely of insects, they probably do more good than harm.

Mus sylvaticus, *Linn.* WOOD MOUSE.

Common everywhere in Essex. A gentle little creature, and a delightful home pet, but one of the most destructive of its race in fields, gardens, or plantations. Plots of newly-sown peas or corn are especial objects of its attention. It is rarely found in houses, barns, or ricks, much preferring the shelter of a hedgerow or wood.

The colour of different specimens varies considerably in its range of shades of red. Albinos are occasionally taken.

Of all our native Mice, this is the most easily tamed. An occasional specimen is more than usually friendly, and may be induced to come into the hand within a month of capture. I generally have some of these little creatures in confinement. They are extremely friendly one with another, and a large number, even of complete strangers, may be kept together. This gregarious way of living seems natural to the Wood Mouse, for fourteen or fifteen, and even more, may sometimes be dug out of one burrow. They work in company in storing provisions, bunches of growing barley or other corn often showing clearly where their storehouse has been.

Nothing in the way of vegetable food seems to come amiss to this very abundant mouse.

Mr. J. E. Harting gives (*Trans. Essex Field Club*, vol. i., p. 92), a quotation from Joshua Childrey's *Britannia Baconia* (London, 1660, 8vo, p. 100), where he says that in 1580 an extraordinary swarm of Field Mice appeared in Dengie hundred, in Essex, which ate up all the roots of the grass. It may be questioned whether the chief offender in committing these depredations was not the Short-tailed Field Vole (*Arvicola agrestris*), under which further mention of this plague of mice will be found (see p. 66).

Mus musculus, *Linn.* COMMON MOUSE.

Who does not know this foul-smelling, but nevertheless pretty, little beast? It abounds everywhere, and has followed man to all parts of the world. Houses, buildings, and corn-ricks are its favourite haunts, and it does not occur in this country except in their vicinity. Its original home certainly was not in Britain.

Albinos occasionally occur wild. In February, 1897, nearly a score of white specimens, having pink eyes, were caught in threshing a stack of wheat on Stebbing Ford Farm, Felstead (*Essex County Chronicle*, February 26th, 1897).

Mus rattus, *Linn.* BLACK RAT.

This, our oldest Rat, was abundant before the advent of the Brown Rat, called by Waterton and others the "Hanoverian Rat." It is now almost extinct, although still occurring about the docks and other places in the East-end of London. The specimens thus met with may not be natives. Probably the race is kept up by escapes from the vessels lying in the river in the neighbourhood.

Mr. H. Barclay records (*Zool.*, 1848, p. 616) the trapping of an immature specimen in a dry bank at Leyton. Two others were also taken near the same place.

The Black Rat is easily known from the Hanoverian Rat by the size of its ears, the slenderness and length of its tail, and by the mouth appearing to be so far under the nose.

In habits and feeding, there is much in common between the two species. *Mus rattus*, however, confines itself to the upper parts and roofs of buildings : *Mus decumanus*, to the basements and drains.

Mus decumanus, *Pall.*   HANOVERIAN, BROWN, OR NORWAY
                                  RAT.

This pest, although placed amongst our native animals, did not make its appearance in England until the earlier part of the eighteenth century, when it was doubtless brought hither by merchant vessels from some southern country. Pennant says it came from the East Indies, and he remarks with prophetic intuition: "It has quite extirpated the common kind (*Mus rattus*) wherever it has taken up its residence, and it is to be feared that we shall scarcely find any benefit from the change, the Norway Rat having the same disposition as, but greater abilities for doing mischief than, the common kind." At the time when the name "Norway Rat" was first applied to it, this rat was not known at all in that country. It was called the Hanoverian Rat from its having arrived in this country about the same time as the first Hanoverian sovereign. This, no doubt was a witticism of our Jacobite predecessors.

Its fecundity, cunning, and omnivorous habits enable it to defy all efforts for its extirpation, and the destruction wrought by game-preservers on so-called "vermin," by exterminating its natural enemies, facilitates its continuous abundance in many districts. In the light soils of the neighbourhood of Colchester, every hedge has its colony, especially where the game is strictly preserved.

In many places around the Essex coast, there exist considerable portions of land, sufficiently raised to be above ordinary high tides, but covered periodically with the sea. These spots, called "saltings," are frequently occupied by colonies of the Hanoverian Rat, as are also those detached pieces which form islands, and which are often named "rat islands" from this circumstance. How the occupants of these islands subsist is almost a mystery, for the only vegetation is that common to muddy seashores, and there is no fresh water, even the little pools and puddles being salt. It is possible that the rats may feed on some of the salt plants, on shell-fish and crustaceans, or on animal or fish refuse thrown up by the tide; but it is difficult to conceive how such a thirsty animal finds a substitute for fresh water. There are also many colonies among the stones covering the face of the sea-walls. These will, of consequence, fare better, as there will be within reach plenty of vegetable products suited to their wants, as well as fresh water in any quantity.

Rat colonies may often be found also on the sides of ponds and marsh ditches; for this rat takes to water, swims, and dives almost as freely as the Water Vole. In many instances, where the Water Vole is blamed for the destruction of young ducks, chickens, game, or eggs, the real offender is this waterside Hanoverian Rat.

Few animals fight more desperately for life when driven into a corner. All fear then seems entirely to forsake the ferocious little beast, which appears determined to sell its life as dearly as possible. Few of its natural enemies will attack the Hanoverian Rat under these circumstances. I have never seen cats even attempt it. They prefer to seize the rat whilst running; and in doing this, almost every cat has its own peculiar method. One, I remember, always turned the rat over on its back with her foot, caught it by the throat, and at the same time fell on her side, and gave one violent kick. The result was immediate death to the unfortunate rat. Other cats throw them over their heads after having bitten them through the heart; others, again, simply hold them in their mouths until the rats are dead. But, in whatsoever way the cat seizes them, she always takes care to avoid their formidable incisor teeth, while at the same time she drives her canines into some vital spot. The town rat is very cautious, and most difficult to trap. His country cousin quickly takes warning after a few have been caught, and most adroitly avoids all snares and gins in the future.

Dr. Bree has recorded (*Field*, Oct. 5th, 1872, p. 328) the capture of two hairless Rats at Thorpe-le-Soken. They were entirely without hair, except for their whiskers. He forwarded the specimens to the Museum of the Royal College of Surgeons.

Family ARVICOLIDÆ.

Genus ARVICOLA, *Lacep.*

Arvicola amphibius, *Desmar.* WATER RAT OR WATER VOLE.

This little animal is common in all parts of Essex, wherever there are sluggish streams, or where stagnant water exists in sufficient quantity to hide it.

Bell, quoting G. R. Waterhouse, says: "The animals comprising this family (Arvicolidæ) have all the essential characteristics of the Muridæ, but differ in having rootless molars and in the form of the lower jaw." There are other characters given, but the rootless molars seem to be the great and easily-distinguishing feature separating Muridæ and Arvicolidæ.

It will be well to bear in mind that the common Water Rat is sometimes quite black in colour, and has been described by Macgillivray (*Hist. of Brit. Quad.*, being vol. xiii. of the *Naturalists' Lib.*) under the name of *Arvicola ater*. This is, however, merely a variety. This form has been occasionally mistaken for the old English Black Rat (*Mus rattus*), and many of the supposed appearances of the latter animal can thus be explained.

Its size and numbers considered, the Water Rat does less damage than any other British member of the family. Its principal food is aquatic vegetables, of which it only takes what can well be spared. Occasionally, when it is abundant and the weather severe, it does mischief among osier beds; but the injury inflicted on the farmer is so small as hardly to be worth consideration.

Arvicola agrestis, *Linn.*   COMMON, OR SHORT-TAILED, FIELD VOLE.

This Vole abounds sometimes to such an extent as to entirely destroy the herbage. From the quantity of food consumed by it, it is easy to conceive the devastation it may cause when existing in large numbers. Not satisfied with herbage only, Voles, according to Bell (*Brit. Quadrupeds*, p. 325), were known, many years since, to destroy the plantation of young oaks in the New Forest and in the Forest of Dean. Some years ago, I recorded (*Zool.*, 1881, p. 461) the weight of food, amounting to six drachms (apothecaries weight) of clover, consumed during the space of four-and-twenty hours, by a specimen I had in confinement.

Its insatiable appetite compels it to be abroad at all seasons of the year and all hours of the day; but I have noticed those in captivity to be more active towards and during the evening. They appear rather stupid, and I never succeeded in making them very tame.

Bell says *Arvicola agrestis* may always be distinguished by the character of its second upper molar, which has five cemental spaces, whereas the tooth in *Arvicola arvalis* (not yet found in Great Britain), as in all the other European Voles, presents four spaces.

The nest is usually placed among the roots of grass, sometimes under fallen timber. The young are from four to six in number, and there are generally three or four broods in a year. Weasels, Owls, and Kestrels, are their greatest enemies. The Short-eared Owl (*Otus brachyotos*) is a great destroyer of them.

In another part of this work (see under *Mus sylvaticus*, p. 60), I have referred to plagues of mice in Essex.

This subject has received attention from Mr. Miller Christy (*Birds of Essex*, 1890, p. 157), and has been made the theme of a paper by Mr. E. A. Fitch (*Essex Nat.*, vol. iii., p. 178), wherein several interesting quotations from old authors, respecting the depredations of the Short-tailed Vole, are given.

Holinshed says (*Chronicles*, [1586], vol. iii., p. 1315):

"About Hallowtide last past [1580], in the marishes of Danesie Hundred, in a place called Southminster, in the Countie of Essex, a strange thing happened : there suddenlie appeared an infinite multitude of mice, which, overwhelming the whole earth in the said marishes, did sheare and gnaw the grasse by the roots, spoiling and tainting the same with their venemous teeth, in such sort that the cattell which grased thereon were smitten with a murreine, and died thereof; which vermine by policie of man could not be destroied, till now at the last it came to passe that there flocked together all about the same marishes such a number of owles as all the shire was not able to yeeld; whereby the marsh holders were shortlie deliuered from the vexation of the said mice."

Stow, in his *Annales* (1605, p. 1166), makes the same statement, evidently from the same authority. Speed also relates (*Theatre of the Empire of Great Britiane*, 1676, fo., p. 31) the same occurrences, with his own quaint moralising upon it :

"But lest we should exceed measure in commending, or the people repose their trust in the soile; behold what God can doe to frustrate both in a moment, and that by His meanest creatures: for in our age and remembrance, the year of Christ, 1581, an Army of *Mice* so over-ran the marshes in Dengey Hundred, neer unto South-minster in this County, that they shore the grass to the very roots, and so tainted the same with their venomous teeth, that a great Murrain fell upon the Cattel which grased thereon, to the great loss of their owners."

Fuller, in his *Worthies* (1662, p. 348), refers to a devastation of some sixty or seventy years later, but which had a like dire result. He says:

"I wish the sad casualties may never return which lately have happened in this county: the one, 1581, in the Hundred of Dengy (Stow Chron., *Anno Citat.*), the other 1648, in the Hundred of Rochford and Isle of Foulness (rented in part by two of my credible parishioners, who attested it, having paid dear for the truth thereof), when an Army of Mice, nesting in Ant-hills, as Conies in Burroughs, shaved off the grass at the bare roots, which, withering to dung, was infectious to Cattle. The March following, numberless flocks of Owls from all parts flew thither and destroyed them, which otherwise had ruined the country if continuing another year."

William Lilly, the astrologer, in his almanac *Merlinus Anglicus Junior*, alludes to another invasion at Southminster, and says that at the same time in Norfolk, over as much as an hundred acres together, one could hardly set down his foot without treading on them.

Arvicola glareolus, *Schreber.* RED FIELD-VOLE OR BANK VOLE.

This Vole is by no means common in Essex, according to my experience, although the first specimen recorded as British was described by Yarrell from an Essex example (*cf. Proc. Zool. Soc.*, 1832, p. 109).

I have seen a specimen from West Bergholt and another from Layer de la Haye. Probably more might be found if observers were careful to examine all the Voles they met with.

There is a good description of this species by Mr. J. E. Harting (*Zool.*, 1887, p. 365), who records captures at West Bergholt and Layer de la Haye, both in Essex, and gives particulars of its distribution in Great Britain.

Mr. Edward Rosling has recorded (*Zool.*, 1885, p. 433) the capture of an albino specimen near Chelmsford, in

August, 1885. The Vole was rescued by him, uninjured, from a cat, and was forwarded alive to the Zoological Gardens.

The habits of *Arvicola glareolus* appear to be similar to those of the last species, but I think it is never found in such damp situations as the Field Vole. The only distinguishing characters to be entirely depended upon are the teeth. Colour, length of tail, and brush at the end of it, are uncertain marks in such a variable family, and I advise no one to trust any of these traits singly in the identification of specimens.

Family LEPORIDÆ.

Genus LEPUS, *Linn.*

Lepus timidus, *Linn.* COMMON HARE.

I shall say little about this animal, as it must be well-known to everyone. It occurs in all parts of the county, and is, from its manner of feeding, a great pest to the corn-grower and the gardener.

Hares vary much in weight. From seven to eight pounds is the average in this county; but I once saw a female turn the scale at ten pounds and a half.

Occasionally a perfectly black Hare has been captured. Mr. James Cooper records (*Zool.*, 1856, p. 5058) shooting one such on the estate of Sir W. Bowyer Smijth, Hill Hall, Epping, on January 31st, 1856. Others are noted from Ongar (*Field*, Jan. 3rd, 1863, p. 16), and Epping, where within a circuit of about five miles four black Hares and leverets were taken about the year 1865.

I have once seen a very light-coloured, almost white, Hare at Paglesham. A similar specimen was killed, after a very long chase, on the estate of Sir C. C. Smith, at Suttons, near Romford, in November, 1865 (*Field*, Nov. 25th, 1865, p. 377).

In the *Supplement* to Daniel's *Rural Sports* (London, 1813, 4to, p. 480), there is an account of a Hare being run down by a single dog, a cross between a Hound and a Spaniel—and that a nearly blind one—at Great Baddow in December, 1808. The chase was for a wager. The same author gives (vol. i., p. 548) the details of a very long course near Felstead, in February, 1789. Both occurrences are of some interest, as showing the amount of sport afforded by this creature.

Lepus cuniculus, *Linn*. RABBIT.

This destructive creature is very common in all parts of Essex, and little need be said about it.

Its food, habits, and appearance are sufficiently well-known to all dwellers in the country.

It may not, however, be so well-known, that occasionally specimens occur, which in colour are quite black, or white, and this, I believe, without any admixture of tame blood.

Mr. Reginald W. Christy gives (*Essex Nat.*, vol. ii., p. 33) a description of several black Rabbits that he observed at Roxwell. Mr. E. A. Fitch describes (*ibid.*, vol. iii., p. 25) a litter of albinos found at Hazeleigh. In neither instance does there seem any reason to suspect a cross with a domesticated parent.

There are, in the large warrens of the counties of Norfolk and Suffolk, any number of silver-grey rabbits. I do not hear of any of this variety in Essex, unless the black ones mentioned by Mr. R. W. Christy, may be sports in this direction, for the young silver-grey animal is generally black in its first coat.

Order RUMINANTIA.

Family CERVIDÆ.

Genus CERVUS, *Linn.*

Cervus elaphas, *Linn.* RED DEER or STAG.

About 1827, according to Mr. J. E. Harting (*Trans. Essex Field Club*, vol. i., p. 79) and Mr. E. N. Buxton (*Epping Forest*, 1897, p. 62), the last Red Deer were removed from Epping Forest to Windsor. Until that date, this species had continuously from the earliest times been a resident, in a wild condition, in this county, as the various mentions of Red Deer in the Forest Records attest. Under the Forest law of Canute, deer and boars are among the principal beasts for the killing of which recompense must be made.

From the time of Edward the Confessor, the Kings have hunted in Epping Forest. It is said (Nott's ed. of *Surrey's Works*, i., xxxiv.) that Henry VIII. was actually enjoying the sports of the Forest when his Queen Ann Boleyn was executed. A pre-arranged signal, the report of a distant gun, informed him of her death. Queen Elizabeth was very fond of shooting a buck with the cross-bow, and, in May, 1578, she was entertained for several days by the Earl of Leicester at Wanstead House. In Queen Elizabeth's Lodge, at Chingford, we have a record of her visits to the Forest.

After the Restoration, £1,000 was expended in restocking the Forest with Deer (*Cal. State Pap. Dom. Chas. II.*), which during the Commonwealth were evidently neglected.

Mr. Harting says (*Essex Naturalist*, vol. i., p. 54), that, early in last century, the Deer of Epping Forest decreased largely in numbers, owing partly to the demands of claims made for "fee deer," partly (and this he quotes from Gilbert White's letter to Pennant, *Selborne*, ed. 1875, pp. 22-23), to the depredations of the gang known as "Waltham

UNIV. OF
CALIFORNIA

Laver's "*Mammals, etc., of Essex.*" [*To face p.* 71.

Haunts of Deer in Epping Thicks.

Blacks," who surreptitiously carried off large numbers. It is obvious, however, that White's reference is to Waltham Chase, a district in Hampshire, on the borders of the New Forest. It appears by the Court Rolls of the date that an order was made to the effect that the stock of Red and Fallow Deer in Waltham Chase being so low that they were likely to be extirpated, no more were to be taken for three years.

In the early part of the eighteenth century, the Royal buckhounds hunted the Essex Deer. The Treasury Records show that in 1729 his Majesty's Hounds killed thirteen stags, and in 1730, nine. Later, the Deer in the Forest were hunted by Mr. Tylney Long Pole Wellesley, who kept his pack of old-fashioned staghounds at Wanstead House, and dressed his servants in Lincoln green. A writer in *The Sporting Magazine* for April, 1809, states that Mr. Wellesley's hunt was called the Ladies' Hunt, because so many ladies of the neighbourhood joined in it. The meetings were generally at Fencepiece in Hainault Forest, and upon Easter Monday was held the anniversary meeting, ending with a dinner and a ball. To this, it was customary for many Londoners to resort. Some, of course, were invited guests : others were merely strangers and lookers on, who had come to enjoy the holiday sports. During the mastership of Joseph Mellish, Wellesley's predecessor, although the ball and entertainments had not then been instituted, the annual Easter Hunt was kept up, and obtained great notoriety as an outing for cockney sportsmen. It was ridiculed in drawings, and on the stage ; verses were published in *The Sportsman's Vocal Cabinet*; and, finally, Hood, during his residence at Wanstead (1832-4), immortalised it in his well-known and witty poem of " the Epping Hunt," in which he celebrates the little town, justly famed for butter, and sausages,

"But famous more, as annals tell,
 Because of Easter chase,
Where every year, 'twixt dog and deer,
 There is a gallant race."

The famous Epping Forest Easter Hunt was supposed by Harrison Ainsworth (see his novel *The Lord Mayor of London*) to owe its origin to the sporting habits of the Lord Mayors of olden days.

ANTLERS OF THE LAST RED DEER FROM HAINAULT FOREST.

More information about Essex Red Deer is conveyed in *The Field* (April 5th, 1884), and in Messrs. Ball and Gilbey's *Essex Foxhounds* (1896, pp. 308-335). There are few records relating to the county but make mention of the important part this animal has played in the history of our various families and estates.

Mr. J. E. Harting quotes (*Essex Nat*, vol. i., p. 55) from a manuscript note (by Cary himself) in a copy of Cary's *Survey of the Country Fifteen Miles round London* (1786), owned by Mr. B. G. Cole, the statement that the Crown has an unlimited right to keep Deer in these forests, of which, during Cary's time, as in Norden's (*Description of Essex*, 1594, ed. 1840, p. 9), there was a goodly stock, both of Red and Fallow Deer. In another manuscript note, Cary records: "1827, Oct. 20th. I met the staghounds at Hoghill

House, in Hainhault Forest, to unharbour a stag. After drawing the coverts for a short time, a fine old stag was roused and took a turn round the Forest away for Packnall Corner, hence to Dagenham, and was taken at Plaistow." He adds, "Red Deer to be so near the metropolis, in their wild state, I consider as a singular circumstance."

Above is a sketch, by Mr. H. A. Cole, of a pair of antlers of the Red Deer, now in Mr. W. Cole's possession, and said to have belonged to the last wild Red Deer killed in Hainault Forest, early in the present century (see *Essex Nat.*, vol. i., p. 56, *foot-note*).

Turning to other parts of the county, we find that Sir John Bramston, of Skreens, Roxwell, gives, in his *Autobiography* (Camden Soc., London, 1845, 4to.), an account of a visit paid by James II. to the Duke of Albemarle at New Hall, Boreham, for the purpose of hunting the Red Deer. On May 3rd, 1686, the king arrived in the neighbourhood, and, hearing that the Duke was with the hounds near Bicknacre Mill, he turned in that direction and pursued the chase almost as far as Wanstead, the quarry being at last killed between Romford and Brentwood. The king was near at the death, and did not reach New Hall until nine o'clock at night. Nevertheless, he hunted again next day with renewed vigour, and this time the Deer ran by Broomfield and Pleshey to the Roothings, and was killed at Hatfield Broad Oak. Sir John's description of the Roothing country will answer also to-day: "His Majestie [he says] kept pretie neere the doggs, though the ditches were broad and deep, the hedges high, and the way and feilds dirtie and deepe." (*cf.* also *Essex Nat.*, vol. iii., p. 193.)

Fortunately, however, we can still say we have wild Red Deer in the county, for within a few years some have been brought back from Windsor to re-stock Epping Forest (see W. R. Fisher's *Forest of Essex*, London, 1887, 4to, p. 220). Professor Flower records (*Zool.*, 1887, p. 344) the existence of a small herd in Takeley Forest, near Hatfield Broad Oak, the progeny of a single hind lost by the hounds during a chase. A writer in the *Zoologist* (1888, p. 74) testifies that both Red and Fallow Deer are still existing in Epping Forest.

We may therefore with justice add this species to our list of the Fauna of Essex, since, with the exception of about fifty years of the present century, Essex has never been without wild Red Deer.

The curious variation of names for the young of the Red Deer, of both sexes, is the subject of an interesting paragraph in Fisher's *Forest of Essex* (p. 193); and Mr. Harting has described (*Trans. Essex Field Club*, vol. i., p. 80) the growth of antlers which determines the names. Thus the animal is known by the names of a Calf, a Brocket, a Spayad, a Staggard, the fifth year a Stag, and in and after the sixth year, a Hart. But, if he had been hunted by the King he became a Hart Royal, and, if the King, in consideration of the sport given, had proclaimed he was not to be hunted again, he became a Hart Royal Proclaimed. The female, called first a Calf, next a Herst, from the third year onwards is a Hind.

Some information as to the number of Red Deer now existing in Essex Deer Parks is given under the heading Fallow Deer (*infra*, p. 76).

Cervus dama, *Linn.* FALLOW DEER.

This animal is probably an introduction into Britain, of which it is not therefore a true native; but, as it exists in many parks in this county in a semi-domesticated condition, and has been for many centuries truly feral in Epping Forest, we may fairly claim it as an Essex animal. It is by some supposed to have been introduced by the Romans.

UNIV. OF
CALIFORNIA

Laver's "Mammals, etc., of Essex.'] [*To face p.* 75

Head of Fallow Buck, from the Weald Hall Herd.

Although fossil remains of the Red and Roe Deer are not infrequently discovered, none of the Fallow Deer have yet come to light.

The judicial decision which fortunately placed Epping Forest under the charge of the Corporation of London came just in time to save the remnant of the Epping Fallow Deer, as in 1870 the stock in the Forest had dwindled down to only five or six brace of deer and one buck (Fisher's *Forest of Essex*, p. 221, *The Field*, August 5th, 1876, p. 156, and *Zool.*, 1888, p. 74).

We have already referred to Mr. J. E. Harting's exhaustive paper (*Essex Nat.*, vol. i., p. 46) on "The Deer of Epping Forest," which is accompanied by illustrations. In pointing out some of the peculiarities of the Epping Fallow Deer, he says (p. 56): "The Fallow Deer have held their own, in spite of all difficulties, until the present time, and have strangely preserved their ancient character in regard to size and colour." He goes on to describe them as comparatively small in size, of a uniform dark brown, almost black, colour, in which respect they vary from herds in other parts of the country, and with very attenuated antlers—characters which he considers show, by their persistency, the probable antiquity of the stock. Mr. E. N. Buxton, a Verderer of the Forest, considers them to be the only representatives in England of the ancient Deer (*Epping Forest*, 4th ed., Lond., 1897, p. 58). A drawing of the head of a Fallow Buck from the Forest (now in the Epping Forest Museum), sketched by Mr. H. A. Cole, forms a frontispiece to the present work. It contrasts strongly with the head of another Fallow Buck, belonging to the Weald Hall Herd, and also sketched by Mr. H. A. Cole, which faces this page. This also is in the Epping Forest Museum.

The list of parks in Essex containing Deer in the year 1892 was given in *The Essex Review* (vol. iii., 1894, p. 136), and may well be appended here. It is derived from *A Descriptive List of the Deer Parks and Paddocks of England*, by Joseph Whitaker, F.Z.S. (London, 1892). The number of deer parks appears to be ten, about which there is the following information:

EASTON PARK.—700 acres, 450 Fallow Deer, 120 Red Deer. Well-timbered and well-watered; many fine oaks; rather flat.

HATFIELD FOREST.—500 acres, 300 Fallow Deer, 10 Red Deer. Flat; well-timbered; some enclosed game coverts in the park; ponds; rather wild; oaks very fine. It was till recent times a forest.

THORNDON HALL.—North Park, 341 acres: South Park, 373 acres; 50 Fallow Deer, 40 Red Deer. Timber very fine; park undulating; scenery varied and picturesque. A large herd of deer, about 1,200 strong was killed down some few years ago, after the destruction by fire of the mansion.

WEALD PARK.—300 acres, 80 Fallow Deer, 70 Red Deer, 9 Japanese Deer, 2 Roe Deer. Very fine oak and hornbeam timber; large amount of fern, five to six feet high in places.

BOREHAM PARK.—300 acres, 120 Fallow Deer.

BELHUS PARK.—300 acres, 100 Fallow Deer (formerly 300). This park has the ancient and now uncommon right of "free warren."

MARKS HALL PARK.—200 acres, 200 Fallow Deer.

WYVENHOE PARK.—180 acres, 100 Fallow Deer (all black). Fine old oaks (between four and five hundred years old), limes, elms, and beeches; eighty acres of fern; four acres of ornamental water and a brook intersect the park, forming the boundary between Greenstead and Wyvenhoe parishes.

LANGLEY'S PARK.—100 acres, 88 Fallow Deer.

QUENDON PARK.—80 or 90 acres, about 100 Fallow Deer. Some very fine oaks. Has been a deer park for about two hundred years.

Mr. Evelyn P. Shirley, in his *English Deer Parks* (London, 1867), described three more parks—Audley End, Short Grove, and Braxted—which now no longer contain Deer.

Genus CAPREOLUS, *H. Smith.*

Capreolus caprea, *Gray.* ROE DEER.

Again I must quote Mr. J. E. Harting's valuable paper on "The Deer of Epping Forest" (*Essex Nat.*, vol. i., p. 58) for an account of the Roe as an Essex animal. In it, he shows conclusively from charters, court rolls, and other satisfactory proofs, some of them geological, that the Roe was formerly an inhabitant of the county. It disappeared from the Forest of Essex, apparently before Norden wrote his *Description of Essex* in 1594. Mr. Harting also details the active part he took, in company with Mr. E. N. Buxton, one of the verderers of Epping Forest, in successfully reintroducing to the Forest, in 1884, this interesting and beautiful creature (see also *Field*, April 5th, 1884, pp. 487-8). By this most enterprising restoration, we are enabled to add another species to our Fauna. The Roes are now (1897) doing well, and are supposed to number over twenty.

In excavating the remains of a Roman building at West Mersea in the spring of 1897, bones and antlers of the Roe Deer were found, with those of the Sheep and the small Celtic Ox.

Order CETACEA.

Sub-order MYSTACOCETI.

Family BALÆNOPTERIDÆ.

Genus BALÆNOPTERA, *Lacépède.*

Balænoptera musculus, *Linn.* RORQUAL.

This whale, one of the largest animals extant, has several times occurred on our coast.

One was taken in the Thames in May, 1859 (Bell's *Brit. Quad.*, 2nd ed., p. 400).

The capture of a "finner whale" at Grays was previously noted (*Zoologist*, 1849, p. 2620); and, judging from the dimensions given (viz., length 58 feet, girth 30 feet), it was probably an example of this species. Whatever question there may be as to identification in this case, there can be none with reference to that captured at Burnham on February 12th, 1891. This very interesting specimen is fully described by Mr. Walter Crouch (*Essex Nat.*, vol. v., p. 124), who gives a table of twenty-two different measurements of its huge bulk. It was remarkable in showing a considerable surface about the jaws which was perfectly white. In other portions of the body also, the same colour was unusually developed. The article is further illustrated by Mr. Crouch with a drawing by himself of the Whale, which drawing is herein reproduced.

Mr. E. A. Fitch records (*Essex Nat.*, vol. v., p. 134) the appearance of another specimen in the Crouch a few weeks later, in April, 1891. From the description and estimated size, this was probably of the same species.

Balænoptera borealis (laticeps), *F. E. Gray.* RUDOLPHI'S RORQUAL.

This Whale, which is rare on the British coasts, has been four times captured in Essex waters within the last few years.

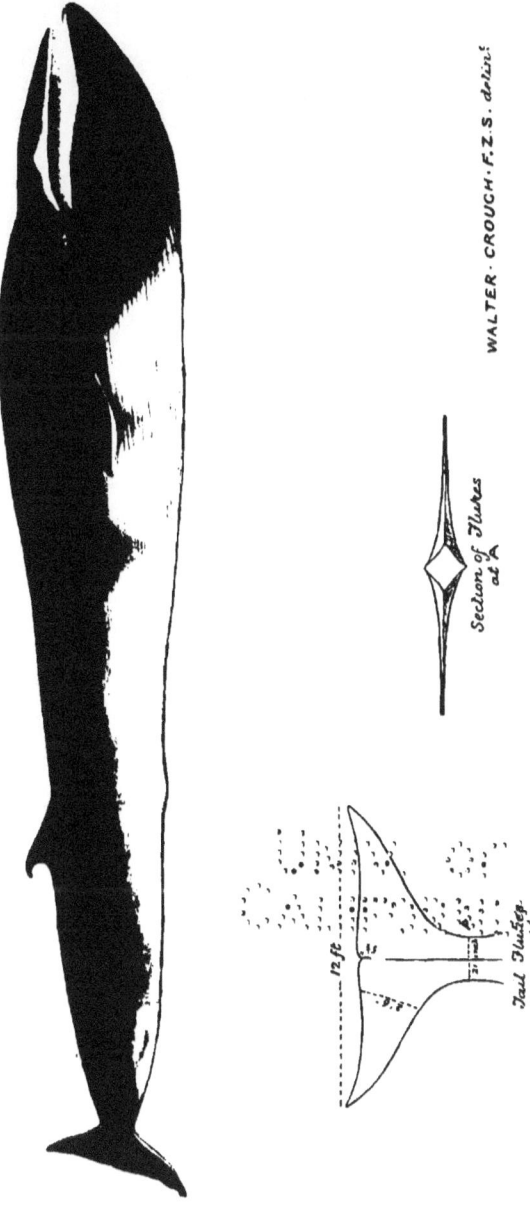

The Common Rorqual (*Balænoptera musculus*) captured in the River Crouch, Feb. 12, 1891.

The first record I can find of the species is in a paper by Dr. J. E. Gray (*Proc. Zool. Soc.*, 1864, p. 218), where he mentions this rare Rorqual being found at Hope Reach, in the Thames, near Gravesend, in the year 1859. The second was stranded near Crixea, in the river Crouch, on the 8th November, 1883. It was identified by Professor Flower and described by him (*Proc. Zool. Soc.*, 1883, p. 514 ; and *Trans. Essex Field Club*, vol. iv., p. 3). The same example is mentioned by Mr. A. H. Cocks (*Zool.*, 1886, p. 129.) The third was found dead at Tilbury, and was identified, described, and drawn by Mr. Walter Crouch (*Essex Nat.*, vol. ii., p. 41). The fourth was captured in the Medway, and, having passed through the Thames estuary, must have been in Essex waters. We can, therefore, claim it as an Essex specimen. This also was identified and described by Mr. Walter Crouch in the *Rochester Naturalist* for 1888, where a figure and sixteen measurements are given (*cf. Zool.*, 1888, p. 466).

Balænoptera rostrata, *Fabricius.* LESSER RORQUAL.

This is one of the best marked and most easily distinguished species of the family, and at the same time one of the most common on our coasts. It has occurred in the Thames several times. John Hunter, the famous anatomist, describes in the *Philosophical Transactions* (1787, p. 448), one caught upon the Dogger Bank, and afterwards purchased by himself. Another, also recorded and figured (*Zoologist*, 1843, p. 33), is now preserved in the British Museum.

Sub-order ODONTOCETI.
Family PHYSETERIDÆ.

Genus PHYSETER, *Linn.*

Physeter macrocephalus, *Linn.* SPERM WHALE.

This tropical Whale has occasionally wandered to the shores of our island. It is recorded in Bell's *British Quadrupeds* (2nd ed., p. 417) that, in 1788, six were found dead on the Kentish coast. A live one ran ashore in the Thames at the same time. Dale (*Hist. of Harwich*, 2nd ed., 1732, p. 413) also mentions one caught in the Thames, and brought ashore at Blackwall.

An original manuscript letter from Walberswick, Suffolk, dated 7th March, 1788, preserved in the British Museum copy of the volume of the *Philosophical Transactions* for 1787, records the appearance of twelve Sperm Whales, after a hard gale of northerly winds, in February, 1763. Two of these were driven ashore, dead, on the coast of Essex—the writer does not say at what points. He, however, cut up more than one out of the twelve, and gives the dimensions of some of the animals.

Sub-family ZIPHIINÆ.

Genus HYPEROODON, *Lacépède*.

Hyperoodon rostratus, *Chemnitz.* COMMON BEAKED WHALE.

This Whale appears to come into British waters regularly in the autumn, and specimens are killed almost every year on some parts of the coasts of this island. John Hunter records one captured in the Thames above London Bridge in 1783 (Bell's *Brit. Quad.*, 2nd ed., p. 423). Another, no doubt of this species, is figured in Dale's *History of Harwich and Dovercourt* (p. 412), it having been captured in the Blackwater estuary.

In July, 1891, two male specimens of this whale occurred in the Thames (*Essex Nat.*, vol. v., p. 170). One was near the Nore lightship, and was towed into Leigh. The other was found near the entrance to Barking Creek. Dr. Murie identified the Leigh example and Mr. W. Crouch the one near Barking (see *Essex County Chronicle*, Aug. 14th, 1891).

This is the best-known species of the family of ziphioid whales, which are distinguished by possessing one or two pairs of teeth, situated in the lower jaw only.

Family DELPHINIDÆ.

Genus ORCA, *Gray.*

Orca gladiator, *Lacépède.* GRAMPUS.

John Hunter, the anatomist, records the capture of three specimens of this savage and destructive animal in the River Thames, towards the end of the last century (see Bell's *Brit. Quad.*, p. 446). There is, in the British Museum, the skull of a specimen taken on the Essex coast, as recorded in the *Zoologist* for 1873, p. 3429, and Dale (*Hist. of Harwich*, p. 412) mentions another specimen. Some years since, I saw two whales which had been killed in one of the creeks of the Blackwater. These, I have no doubt, were of this species, but no record was kept of them, and I do not recollect what became of their bones. Probably they went, as usual, for manure.

Genus GRAMPUS, *Gray.*

Grampus griseus, *G. Cuvier.* RISSO'S GRAMPUS.

One of this species was found stranded in the Crouch about the 5th September, 1885, just above the spot where the specimen of Rudolphi's Rorqual, previously mentioned, was captured. It was, unfortunately, boiled down before I heard of it. I obtained, however, the remains of the skull and lower jaw. These are now deposited in the British Museum. Professor Flower, after having examined them, confirmed my identification (*Zool.*, 1888, p. 260; *Essex Nat.*, vol. ii., p. 72). This last-mentioned was only the fifth specimen then recorded on the British coasts, and the Crouch River appears to be the farthest point east and north at which this species of Grampus has ever been found.

Genus PHOCÆNA, *Cuvier*.

Phocæna communis, *F. Cuvier*. PORPOISE.

Who does not know this merry and active creature? It occurs everywhere on our coasts, and is as frequently seen during stormy weather as at any other time, apparently revelling in the tempestuous waters. It is so common that I have not thought it necessary to give any records of capture.

Genus DELPHINUS, *Linn.*

Delphinus tursio, *Fabricius*. BOTTLE-NOSED DOLPHIN.

This animal, generally reputed rare, has occurred in Essex several times within my own observation. A female, ten feet long, was captured a few years since off Harwich. Another was shot in the Colne on September 5th, 1881 (*Zool.*, 1881, p. 419), while two more were also taken in the same river during the following year (*Zool.*, 1882, pp. 147, 351). In 1829, one was taken in the Thames below the Nore, and its skeleton is now in the museum of the Royal College of Surgeons (Bell's *Brit. Quad.*, 2nd ed., p. 468).

All these examples were females; but, on May 29th, 1892, I had the opportunity of examining three specimens, two males and one female, which were captured in Mill Creek, Fingringhoe. The female, which was the finest of the three, and which measured ten feet three inches in length, was in full milk. A drawing by Major Bale, of Colchester, of the locality where these specimens were captured, faces this page.

Beside those recorded above, I have seen several other specimen captured in various creeks on the coast. When at sea, also, I have frequently been close enough to recognise the species, as they passed or rested near.

Laver's "Mammals, etc., of Essex." [To face p. 82.

Mill Creek, Fingringhoe, Essex.

In most of the drawings of the skull of this animal, the teeth are represented as truncated. This is no doubt the result of wear, and is incidental to age. In all the individuals examined by myself, the teeth were sharp, and slightly directed inwards and backwards, except one of the males caught at Fingringhoe, which had the truncated teeth, before alluded to as indicative of age. I had the opportunity of examining the stomach of one captured at Harwich. It was nearly empty, but contained some of the ear-bones (otoliths) of the Gadidæ. I recognised those of the cod, the haddock, and what appeared to be those of the whiting.

The species must be easy to kill, as this one was caught by a cod-hook in the lip. Some of the others mentioned above were destroyed by a charge of small shot.

Delphinus albirostris, *Gray.* WHITE-BEAKED DOLPHIN.

As I have elsewhere recorded (*Zool.*, 1889, p. 382, and *Essex Nat.*, vol. iii., p. 169), a school, consisting of seven, or possibly nine, individuals of this rare cetacean, was seen in the Colne, on the 11th of September, 1889, when five of them were captured. Some were shot with rifles, while one was driven aground and killed by a sailor with his pocket-knife. The remarkable white beak and sides attracted the attention of the Colne fishermen; but I could not learn that any one of them had ever before seen a similarly-marked specimen. I therefore consider this species to be very rare on our coast.

[A view of the Estuary of the Blackwater, as seen from West Mersea, in the neighbourhood of which not a few cetaceans have, from time to time, been driven ashore, appears facing page 90. It is drawn by Major Bale, of Colchester].

Class REPTILIA.

Order LACERTILIA.

Family LACERTIDÆ.

Genus LACERTA, *Linn.*

Lacerta vivipara, *Jacq.* VIVIPAROUS LIZARD. Locally, "SWIFT."

This active little creature is common in all parts of Essex, especially so on all furze-covered commons or heaths. It may often be found basking in the sun on a gate or rail, fully exposed to the hottest rays, at the warmest period of the year. I have several times seen specimens in which the tail was bifid, possibly the result of an injury. One such is noted (*Zool,* 1861, p. 7514.)

Family ANGUIDÆ.

Genus ANGUIS, *Linn.*

Anguis fragilis, *Linn.* SLOW WORM OR BLIND WORM.

This is quite common everywhere. It is very extraordinary that this perfectly harmless snake-like lizard should be so generally considered venomous by country people, by some of whom it is even more feared than the Viper. I have often been cautioned of the danger I ran in handling such a poisonous creature. I suppose, in time, better knowledge will prevail; but at present its strong resemblance to a snake induces country people almost universally to destroy it.

Order OPHIDIA.

Family COLUBRIDÆ.

Genus TROPIDONOTUS, *Kuhl.*

Tropidonotus natrix, *Kuhl.* COMMON, OR RINGED, SNAKE.

This reptile is fairly common throughout the county, abounding in wooded and marshy districts. Its fondness for water may account for the numbers found in the latter localities. It swims well, and has been known to catch and eat fish. Some years since, an instance was recorded in *The Chelmsford Chronicle* of the capture, in a pond at the Turrets, Rayleigh, of one which had a gold fish in its mouth. The same fact is recorded in *The Field* (June 25th, 1870, p. 536). The following records speak clearly of its abundance in the marshy districts of Essex. Christopher Parsons mentions (*Zool.*, 1845, p. 1027) shooting nine out of a cluster at Shoebury. Mr. E. A. Fitch says (*Essex Nat.*, vol. i., p. 113) that at Saltcote Farm, Goldhanger, in September, about 1882, Mr. Wakelin shot into a black object as large as a football, which he discovered to be formed of intertwined snakes. Out of the mass, he killed twenty individuals at least. At Moon's Farm, Ashingdon (*Essex Nat.*, vol. iii., p. 170), in two days, August 15th and 16th, 1889, three hundred and three snakes were killed out of a heap of "cavings," a local name for the short straws, broken ears, and rough chaff formed in threshing corn.

Family VIPERIDÆ.

Genus VIPERA, *Laur.*

Vipera berus, *Linn.* ADDER OR VIPER.

This is common in woods throughout Essex, but most frequent on the marshes. It lives principally on Field Mice. I do not think it is so plentiful as it used to be before the large hedgerows were reduced in size. In my earlier days, it was not unfrequent for the stockman to report that he had a sheep or a cow bitten by an adder. To the sheep its bite was rarely fatal, and then only when respiration was interfered with by the swelling produced. In the cow, I never knew it to be fatal, but the supply of milk ceased for some days, and considerable constitutional disturbances resulted. Occasionally, during the shooting season, dogs were "stung," but these mostly recovered. Some dogs develop quite a fondness for killing these reptiles, and manage it with perfect impunity. Mr. E. A. Fitch says (*Essex Nat.*, vol. ii., p. 112), that on February 18th, 1887, while removing an old gate-post on Saltcote Farm, Goldhanger, to replace it, seventeen adders and two snakes were found underneath, having chosen, I suppose, cavities as *hybernacula*.

A specimen of the dark form of the Adder ("var δ" of Jenyns, *Man. Brit. Vert. Animals*, p. 298) was recorded by Mr. W. Cole from Loughton, in 1883 (*Journ. Proc. Essex Field Club*, vol. iv., p. x.), and is now in the Epping Forest Museum.

Class BATRACHIA.
Order ECAUDATA.
Family RANIDÆ.

Genus RANA, *Linn.*

Rana temporaria, *Linn.* COMMON FROG.

This well-known animal is extremely common, and generally distributed, especially in the neighbourhood of water.

Rana esculenta, *Linn.* EDIBLE FROG.

This is not a native of our county; but, as Mr. E. Newman points out (*Zool.*, 1848, p. 2268) that it was naturalised by Mr. Doubleday at Epping, it may fairly, I think, be inserted in this list. I have not heard that it has established itself.

Family BUFONIDÆ.

Genus BUFO, *Laur.*

Bufo vulgaris, *Laur.* COMMON TOAD.

Like the frog, this is common and generally distributed.

Order CAUDATA.
Family SALAMANDRIDÆ.

Genus MOLGE, *Merr.*

Molge cristata, *Laur.* GREAT WATER NEWT. Locally, "WATER SWIFT."

Common, and found throughout the county.

Molge vulgaris, *Linn.* COMMON NEWT OR EFT. Locally, "WATER SWIFT."

This is well named the Common Newt, being abundant everywhere in suitable localities, and much more frequent than either of the other species.

Molge palmata, *Schmit.* PALMATED NEWT.

This is not rare. I have found it in several parts of Essex, and abundantly in a pond on Donyland Heath. I have probably observed its occurrence more than I should otherwise have done from the fact of the late Mr. Baker, of Bridgwater, the discoverer of the species as an inhabitant of Britain (*vide* Bell's *British Reptiles*, 1849, p. 155), having shown me, in 1846, some of the first examples he captured.

Mr. G. A. Boulenger, says (*Zool.*, 1886, p. 250) *Molge palmata* is quite abundant in a small pit near Chingford Station, in company with *M. cristata* and *M. vulgaris.* Previous papers by this gentleman had given the distribution of the Palmated Newt in Scotland and England, and it is gratifying to add Essex to the list of English counties where it is found. The record is noted by the editor of the *Essex Naturalist* (vol. i., p. 8), who quotes also from Mr. Boulenger, the distinguishing marks of the two allied species in their winter attire.

Class PISCES.

Sub-Class TELEOSTEI.

Order I. ACANTHOPTERYGII.

Family PERCIDÆ, *Day.*

Genus PERCA, *Artedi.*

Perca fluviatilis, *Rondeletius.* PERCH.

This well-known fish occurs in all parts of the county, our sluggish streams apparently suiting its requirements. One weighing four pounds is reported from Dagenham Lake (*Field*, October 29th, 1881, p. 624).

Genus LABRAX, *Cuvier.*

Labrax lupus, *Cuvier.* BASS or BASSE.

A fish taken occasionally in all our estuaries. Mr. E. A. Fitch records (*Essex Nat.*, vol. ii., p. 19) the capture of several in the Blackwater. They are very good for the table. I have seen some nearly twenty pounds in weight.

Genus ACERINA, *Cuvier.*

Acerina vulgaris, *Cuvier.* RUFF.

In a paper on "Isaac Walton and the River Lea," by Lieutenant R. B. Croft (*Trans. Herts. Nat. Hist. Soc.*, 1881-1883, vol. ii., p. 12), this species is given as an inhabitant of that river. I know of no other Essex stream in which it occurs; but Yarrell says (*British Fishes*, 1836, vol. i., p. 18) that it is found in the Cam, which river rises in Essex.

Family COTTIDÆ, *Linn.*

Genus COTTUS, *Linn.*

Cottus gobio, *Linn.* MILLER'S THUMB.

This occurs plentifully in all our streams, preferring those with a gravelly bottom.

Cottus grœnlandicus, *Cuv.* GREENLAND BULL-HEAD.
(Locally, BULL-ROUT.)

Mr. J. T. Carrington records (*Zool.*, 1880, p. 147) the capture in white-bait nets of several specimens of this fish in the Thames. The identification was confirmed by Dr. Day. I have seen several examples which had been taken on our coast, and have caught them myself in all our estuaries, but have never seen any approaching the size they are said to attain in Greenland.

Cottus scorpius, *Linn.* FATHER LASHER.

This is a very common fish all round the coast. It takes a bait greedily.

Cottus bubalis, *Bloch.* BUBALIS.

Also very common. Both this and the last-named species are frequently caught by persons fishing from piers, even in rather shallow water; and they are constantly captured in shrimp and other trawls. The local name for all three species is "Bull-rout."

Cottus quadricornis, *Linn.* FOUR-HORNED COTTUS.

This is very rare, but Leonard Jenyns says (*British Vertebrate Animals*, Cambridge, 1835, p. 346) some specimens in the British Museum were found among sprats taken at the mouth of the Thames.

Genus TRIGLA, *Artedi.*

Trigla cuculus, *Linn.* RED GURNARD, OR ELLECK.

I have often caught this fish in the estuary of the Blackwater, opposite West Mersea. During those seasons in which it occurs there, it may frequently be captured by means of hooks and by trawling. Sometimes, according to my experience, it appears to be absent for several seasons. I do not consider it worth cooking.

Trigla hirundo, *Linn.* TUB-FISH.

I have also taken this fish in the same estuary, but not so frequently. The pectoral fins of the species are very beautifully coloured.

Trigla gurnardus *Linn.* BLOCH'S, OR THE GREY, GURNARD.

I once, and once only, caught this fish in some numbers, during September, in the Blackwater, off Mersea. In the Appendix to Dale's *History of Harwich* (p. 431), under Grey Gurnard, is the following statement: "This I have seen caught in the sea before this [*i.e.,* Harwich] harbour."

Trigla lyra, *Linn.* PIPER.

Dale (*loc. cit.*) says: "This was caught near Harwich."

Family CATAPHRACTI.

Genus AGONUS, *Bloch.*

Agonus cataphractus, *Bloch.* POGGE.

This curious fish is very frequently taken in the shrimp nets, and small ones may be often seen amongst the boiled shrimps. Yarrell says (*British Fishes*, vol. i., p. 71): "On the eastern coast it is very plentiful."

The Estuary of the Blackwater, from West Mersea.

Family PEDICULATI, *Cuvier*

Genus LOPHIUS, *Artedi.*

Lophius piscatorius, *Linn.* ANGLER. Locally, TOAD-FISH, FISHING FROG, OR SEA DEVIL.

This ungainly fish has been frequently captured on the Essex coast; but it is of no value, and it is more frequently brought on shore as a curiosity than for food. Mr. W. J. Beadel records (*Field*, June 20th, 1863, p. 598) that he took one while trawling in about eight fathoms of water on the edge of the Maplin Sands. Neither he himself nor fifty fishermen who saw it were able to give the fish a name, so it was sent to Mr. Frank Buckland, who soon identified it as *Lophius piscatorius*, or, as he called it, " the Fishing Frog."

Dr. Bree minutely describes (*Field*, March 27th, 1869, p. 260) a specimen caught by some Wyvenhoe fishermen in the Colne, two or three years previous to that date, and which he purchased from them.

The most salient peculiarity of this fish is the totally-disproportionate size of the head and thorax, which, in the last-named example, measured twelve inches, out of a total length of thirty-two inches. This measurement was made from the fish in a dried condition.

Family TRACHINIDÆ, *Risso.*

Genus TRACHINUS, *Cuvier.*

Trachinus draco, *Linn.* GREATER WEEVER.

A fish which is caught occasionally in eel-trawls on the muddy shores. Mr. E. A. Fitch records (*Essex Nat.*, vol. iii., p. 188) the capture of one at Stansgate by "spruling." It is, however, with us of no value except for bait.

Trachinus vipera, *Cuv.* VIPER WEEVER.

This is also taken in the eel-trawls in the same situations as the last, and, like it, is very free in using its formidable spine.

Family SCOMBRIDÆ, *Cuv.*

Genus SCOMBER, *Artedi.*

Scomber scomber, *Linn.* MACKAREL or MACKEREL.

Dale says (*Hist. of Harwich*, 1732, p. 429): "These in their season are here to be caught." Lindsey says (*A Season at Harwich*, London, 1851, pt. ii., p. 83) that, on June 30th, 1821, enormous numbers were caught on the Suffolk coasts. They were of the estimated value of £14,000, sixteen Lowestoft boats alone obtaining a haul of the value of £5,252. Day says (*Fishes of Great Britain and Ireland*, London, 1880-1884, vol. i., p. 90) their habitat extends from the south coasts to those of Suffolk and Norfolk, therefore the Essex coast must be included. I have not myself seen any specimens taken off the coast of Essex.

Genus ORCYNUS, *Lütken.*

Orcynus thynnus, *Lütken.* COMMON TUNNY.

Jenyns (*British Vertebrate Animals*, p. 363) calls it rare; and, quoting Donovan, he states that three were captured in the mouth of the Thames in 1801, and brought to Billingsgate market.

Family STROMATEIDÆ, *Swainson.*

Genus CENTROLOPHUS, *Lacépède.*

Centrolophus pompilus, *Cuv.* BLACK FISH.

I had the pleasure of sending a specimen of this rare fish to Dr. Günther for the British Museum. It was captured in the Colne, and was described by that eminent naturalist (*Ann. and Mag. Nat. Hist.*, vol. ix., 1882, pp. 204 and 338; also in *Zool.*, 1882, pp. 75 and 152).

Family CARANGIDÆ, *Günther*.

Genus CAPROS, *Lacépède.*

Capros aper, *Lacépède.* BOAR FISH.

According to Day (*Fishes of Great Britain and Ireland*, vol. i., p. 137), numbers were caught at Harwich and Southend, about May, 1879. Mr. J. T. Carrington further states (*Zool.*, 1879, p. 342) the fact of the capture of these specimens in a shrimp trawl.

Family CYTTIDÆ, *Kaup*.

Genus ZEUS, *Cuv.*

Zeus faber, *Linn.* JOHN DORY.

This fish is occasionally taken in the shrimp trawls, but all the specimens that I have ever seen have been small, and none of them exceeded seven inches in length.

Family XIPHIIDÆ, *Agassiz*.

Genus XIPHIAS, *Artedi.*

Xiphius gladius, *Linn.* SWORD FISH.

Mr. Montford records (*Zool.*, 1847, p. 1911) the finding of a dead one off the coast of Essex in 1834. C. Parsons also mentions (*Zool.*, 1862, p. 8289) the capture of a living one in Potton Creek, the sword of which was three feet long. An Essex specimen mentioned by Day (*Fishes of Great Britain*, vol. i., p. 148) was probably the same. Buckland (*Famil. Hist. of Brit. Fishes*, p. 37) says a fine specimen, eight feet eight and a-half inches long, of which he made a cast, was caught at Leigh, near Southend, in November, 1866 (see also *Field*, Nov. 3, 1896, p. 362).

Family GOBIIDÆ, *Cuv.*

Genus GOBIUS, *Artedi.*

Gobius ruthensparri, *Retz.* TWO-SPOTTED GOBY.

This little fish is common on the shores of the Wallet, where it flourishes on the hard, almost rock-like, London clay. It is frequently caught and boiled with shrimps.

Gobius minutus, *Donovan.* YELLOW GOBY, ONE-SPOTTED GOBY, and TAIL-SPOTTED GOBY (young).

Common. Day (*Fishes of Great Britain*, vol. i., p. 166) calls it numerous at the mouth of the Thames, and says that it is commonly found amongst the Whitebait brought to the London market. Yarrell says (*British Fishes*, vol. i., p. 260) it is apparently a new species from Colchester.

Genus APHIA, *Risso.*

Aphia pellucida, *Moreau.* SLENDER, OR TRANSPARENT, GOBY.

Possibly this may be common on the Essex coast; but, as the shrimpers throw away all the small unsaleable fish they catch, the opportunities for seeing it are few. The only examples coming under my observation were captured in the Wallet, by the fisherman employed at the Biological Station at Brightlingsea.

Family CALLIONYMIDÆ, *Richardson.*

Genus CALLIONYMUS, *Linn.*

Callionymus lyra, *Linn.* DUSKY SKULPIN. Locally, FOX or DRAGONET.

The shrimp-trawlers very often capture this species in their shrimp nets on our sandy shores, a locality just suited to its habits. Day's figure (*Fishes of Great Britain*, pl. liv.) was taken (*Op. cit.*, vol. i., p. 177) from one captured at Southend, where the reddish specimens are called "Foxes"

(Yarrell, *British Fishes*, vol. i., p. 266). Lindsey says (*A Season at Harwich*, pt. ii., p. 89) this fish is often found in the stomach of the Codfish.

Family DISCOBOLI, *Cuv.*

Genus CYCLOPTERUS, *Linn.*

Cyclopterus lumpus, *Linn.* LUMP FISH.

I have occasionally taken this fish while trawling in the Blackwater. It appears to be frequently caught on the coast, judging by the numbers that one sees exposed for sale. As food, I do not consider it of much value, but tastes differ.

Genus LIPARIS, *Artedi.*

Liparis vulgaris, *Flem.* SEA SNAIL.

This is very common, and frequently taken in shrimp- and eel-trawls. It is a very variable fish, some specimens being beautifully coloured, while others are almost without markings, especially if at all young.

Couch says (*Hist. of Fishes of Brit. Islands*, London, 1860-65, vol. ii., p. 191) it is found even at the mouth of the Thames.

Liparis montagui, *Cuvier.* NETWORK, OR MONTAGU'S SUCKER.

This is not infrequent in shrimp trawls, and Day says (*Fishes of Great Britain*, vol i., p. 187) it is common off the mouth of the Thames.

Family BLENNIDÆ, *Swainson.*

Genus ANARRHICAS, *Artedi.*

Anarrhicas lupus, *Linn.* SEA WOLF.

This fish has been captured at Walton-on-the-Naze (*Essex Standard*, Aug. 29th, 1885). Common as it is on many parts of the English seaboard, it appears only as a straggler off Essex, the sandy shallow shore not suiting its habits.

I have seen many exposed for sale in Norway. As they are there sold alive, the fishermen, to prevent their becoming dangerous, cut off a large part of the upper jaw.

Genus CENTRONOTUS, *Bloch.*

Centronotus gunnellus, *Bloch.* BUTTER FISH.

This is another fish frequently caught in eel-trawls. I find it very commonly on the Zostera-covered ooze at Mersea. Yarrell (*British Fishes*, vol. i., p. 240) says it is found in the mouth of the Thames.

Genus ZOARCES, *Cuv.*

Zoarces viviparus, *Cuv.* VIVIPAROUS BLENNY.

Not a rare fish. It is taken occasionally among sprats by the stow-boat fishermen, also by eel-trawlers.

Family ATHERINIDÆ, *Günther.*

Genus ATHERINA, *Artedi.*

Atherina presbyter, *Jenyns.* SAND SMELT.

This fish is occasionally captured by the Smelt fishermen at the mouth of the Colne. In 1886, I saw many specimens, and an old man engaged in the work told me they were then more numerous than he had ever before known them to be. As a rule, we very rarely see them. They are far inferior in every way to the true Cucumber Smelt.

Family MUGILIDÆ, *Cuv.*

Genus MUGIL, *Artedi.*

Mugil capito, *Cuv.* GREY MULLET.

Yarrell says (*British Fishes*, vol. i., p. 202) "it occurs constantly on the Essex coast." I consider it common in the season all round the coast, entering and passing some distance up our rivers, as far as the limits of salt water, and perhaps even beyond it. It is most wary, and difficult to keep in the net when enclosed. It jumps over the head-rope, and sometimes makes a grand rush with its companions in a body, tearing its way out, unless the net is in good order. If one succeeds in leaping over the head-rope, the whole shoal follows like a flock of sheep.

Mugil chelo, *Cuv.* LESSER GREY MULLET.

This is not as common as the last-mentioned species. Some specimens were, however, taken, in June, 1895, in the Blackwater Estuary.

Family GASTEROSTEIDÆ, *Day.*

Genus GASTEROSTEUS, *Artedi.*

Gasterosteus aculeatus, *Linn.* THREE-SPINED STICKLEBACK.

From its habits and frequency, this is one of the most interesting of the family. It occurs in all situations, in streams, ponds, and (I had almost written) puddles. It is also to be found in pools where fresh and salt water are commingled. I have found this a most variable species, both in size and armature. In some ponds all the individuals are all similar, whilst in other ponds or streams various forms occur. I have never observed which of the named varieties are the most common. The Three-spined Stickleback is a veritable tyrant of the water, and appears to fear few enemies, although he is so small.

Gasterosteus pungitius, *Linn.* TEN-SPINED STICKLEBACK.
TINKER.

This is not nearly so common as the last species, being found more frequently in small streams rather than ponds, although it occasionally occurs in the latter situations. I have never found it in brackish water. It is equally pugnacious, but it never reaches the size of *G. aculeatus*. I think most of our specimens have only nine spines.

Gasterosteus spinachia, *Linn.* FIFTEEN-SPINED STICKLEBACK.

I have found this fish very commonly in trawling for eels amongst the *Zostera marina* on the muddy shores of the Blackwater. Unlike the rest of the family, this species is entirely confined to the sea: otherwise, its habits are very much those of its brother Sticklebacks.

Family LABRIDÆ, *Cuvier*.

Genus LABRUS, *Artedi*.

Labrus maculatus, *Bloch.* BALLAN WRASSE.

I have only seen one specimen of this fish taken on the Essex Coast. It is now at Brightlingsea. Our muddy and sandy shores do not suit the habits of this family.

Genus CRENILABRUS, *Cuvier*.

Crenilabrus melops, *Cuvier.* GOLDSINNY, CORKWING, OR GIBBOUS WRASSE.

The late Dr. Bree describes minutely (*Field*, December 1st, 1866, p. 420) two specimens which were taken on our coast of this—so far as Essex is concerned—rare fish. He says they were quite unknown to our fishermen.

These are the only examples I have heard of.

Order II. ANACANTHINI, *Day*.

Family GADIDÆ, *Cuvier*.

Genus GADUS, *Cuvier*.

Gadus morhua, *Linn.* COD.

The value of this fish as food is well-known. Most of those of large size captured in the Essex estuaries are not in a fit condition for the table, being more or less diseased.

Many young ones, called "Codling," are, however, captured in the finest condition all round the coast by hooks and nets. Dale says (*History of Harwich*, p. 427) : " Cod-fish is to be caught, in the season, before this [Harwich] Harbour."

Gadus macrocephala, *Tiles.* LARGE-HEADED COD.

Dr. Day (*Journ. Linn. Soc.*, vol. xiv., no. 80, p. 689) describes a specimen caught in the mouth of the Thames, at Southend. This is also recorded in the *Zoologist* (1880, p. 26).

Yarrell saw a large one from the mouth of the Thames; but he considered the abnormal size to be due to disease. The local fishermen call this variety "Lord-fish" (*cf.* Day's *Fishes of Great Britain*, vol. i., p. 278).

Gadus æglefinus, *Linn.* HADDOCK.

I have occasionally taken this species in some numbers in the Crouch; but I do not think it is a very common fish on the Essex coast. A large number were observed and taken, off Purfleet, by the officers of the training-ship *Cornwall* in 1879 (*Land and Water*, March 1st, 1879, p. 179). They have also been noted in Dagenham Breach (*ibid.*, March 22nd, 1879, p. 236), by Mr. P. Hood, who communicated his surprise at finding them in fresh water to F. Buckland. The latter's explanation that the water of Dagenham Breach was brackish accounted for the Haddock choosing it as a spawning ground.

Gadus luscus, *Linn.* BIB OR WHITING POUT (Locally, WULE or WHITING WULE).

This is extremely common during the winter, and is known among the Essex fishermen by the name of "Wule." It is delicate eating, but I do not consider it of much value. When freshly caught, the colouring is extremely beautiful; but the brilliancy soon passes off, and the scales are easily detached.

Gadus merlangus, *Linn.* WHITING.

Another well-known and valued member of this family. This fish arrives in our estuaries about the middle or end of September, and gives good sport to fishermen, as it takes the hook freely. In some seasons, it is very abundant.

Gadus pollachius, *Linn.* WHITING POLLACK.

Dale says (*History of Harwich*, 1732, p. 428) that in his time the Whiting Pollack was sometimes caught, and brought to Braintree market with the other varieties of Whiting.

Genus MERLUCCIUS, *Cuvier.*

Merluccius vulgaris, *Cuvier.* HAKE.

Although this fish is so common on the south and west coasts of this island, I never remember seeing more than one specimen captured on the Essex coast. The specimen was sent me by a dealer to name. This, in itself, is a sufficient proof of the rarity of the species in the neighbourhood. No doubt the shallowness of the water prevents their being more common with us. Dale says (*History of Harwich*, p. 429), "Hake is sometimes caught here. When salted and dried, it is called 'Poor Jack.'" Lindsey also mentions (*A Season at Harwich*, p. 73) this fish and its extraordinary voracity.

Genus MOTELLA, *Cuvier.*

Motella mustela, *Nilss.* FIVE-BEARDED ROCK LING.

This is not common, our muddy shores not suiting its habits, but I possess a specimen taken in the Colne. Day (*Fishes of Great Britain*, vol. i., p. 316) gives the mouth of the Thames among its haunts.

CLASS PISCES.

Genus RANICEPS, *Cuvier.*

Raniceps raninus, *Collett.* LESSER FORK-BEARD.

Day records (*Fishes of Great Britain*, vol. i., p. 321) the capture of one by Mr. S. W. Waud, in May, 1858, in the river Crouch.

Family OPHIDIIDÆ, *Müller.*

Genus AMMODYTES, *Artedi.*

Ammodytes lanceolatus, *Le Sauvage.* LARGER LAUNCE OR SAND EEL.

This is found, but not so commonly as the next species.

Ammodytes tobianus, *Linn.* LESSER LAUNCE.

This is common, but I do not think it is ever especially fished for, as is the case in the West of England, where large numbers are caught for bait.

Family PLEURONECTIDÆ, *Risso.*

Genus RHOMBUS, *Cuvier.*

Rhombus maximus, *Cuvier.* TURBOT.

This well-known and valuable fish is taken on all parts of our coast where suitable ground occurs. It is most frequently captured by trawling; but many are caught on the sandy shores of Foulness Island in "kettles": that is, by means of nets arranged in a V-shape, the apex pointing seawards, and with small spaces left between each set of nets, so that the fish, as the tide rises, pass between the nets, and are left stranded in the points, as the tide falls.

Rhombus lævis, *Rondeletius.* BRILL.

This fish is also captured in the same manner, and is not rare.

Genus PLEURONECTES, *Artedi.*

Pleuronectes platessa, *Linn.* PLAICE.

Although it is very common on all parts of the coast, Plaice is said "not to be taken by the hook." While fishing with a very small hook and light tackle, however, I have taken them very frequently, both in the Crouch and the Blackwater. It is a very good-flavoured fish, although rather watery.

Pleuronectes microcephalus, *Donovan.* SMEAR DAB, LEMON DAB, OR MARY SOLE.

This is taken frequently on the coast and in the estuaries, but only by trawling. It is one of the best of this useful and well-flavoured family.

Pleuronectes limanda, *Linn.* DAB.

This fish is taken continually in the season, on all parts of the coast, by nets and hooks. Though generally considered by the London dealers as of no value, it is, when cooked, according to my experience, nearly or quite equal, to the Sole, and, therefore, far better than the Plaice, which (for some reason that I cannot understand) is much preferred by the dealer. It may be because of the smaller size of the Dab. The largest I have ever caught weighed one pound and a half.

Pleuronectes flesus, *Linn.* FLOUNDER.

A fish common everywhere on all parts of the coast, ascending rivers much beyond the tideway, and very frequently seen in the slightly brackish water of the marsh ditches. Flounders sell everywhere, but I cannot say I think them of much value for the table.

Genus SOLEA, *Klein, Cuv.*

Solea vulgaris, *Quensel.* SOLE.

This very common and delicious fish occurs on all our sandy coasts in great, but largely-diminishing, numbers. The trawl is the chief instrument for capturing it, since it very rarely takes a hook.

Order III. PHYSOSTOMI, *Müller*.
Family SALMONIDÆ, *Müller*.

Genus SALMO, *Artedi.*

Salmo salar, *Linn.* SALMON.

Although we have no river in this county that may be called a Salmon-river, the fact that an occasional fish is taken on our coast entitles us to speak of the Salmon as being still truly a member of our Fish Fauna. In former years, before the Thames was poisoned with sewage, it is well-known that Salmon regularly ascended the river. Yarrell says (*British Fishes*, vol. ii., p. 30) the last Thames salmon of which he had a note was taken in June, 1833. A Salmon was taken at Southend, and another in Leigh Bay in 1875 (*Land and Water*, Sept. 25th, 1875, p. 241). They occur now, however, so rarely (and, as it were, so accidentally) that no fishery for them is carried on. Farmer says (*History of Waltham Abbey*, London, 1735, p. 3) that in his time the Lea afforded plenty of fish, including "some salmon." Mr. Harting has called attention (*Essex Nat.*, vol. viii., p. 197) to records of the capture of Salmon in the Lea in 1816, 1825, and 1833, and a record of Salmon at Waltham Abbey in 1820 is printed in the same magazine (vol. ix., p. 227). Another was taken in the Crouch, near Battles Bridge (*Land and Water*, December 10th, 1870, p. 427). One was taken in the mouth of the Blackwater in 1882 (*Field*, July 1st, 1865), and a few are still caught annually in that river.

Salmo trutta, *Linn.* SEA TROUT OR BULL TROUT.

This is occasionally taken on the coast. Some few years since, a fine specimen, weighing four and a half pounds, designated *Salmo eriox*, was captured by Mr. Marriage in the Colne, on the shallows near East Mill, Colchester (*Land and Water* March 30th, 1867, p. 235). The fish was identified by Buckland, who regarded it as a confirmation of his theory that the East Coast rivers would carry Bull Trout, if not Salmon. Indeed, he regarded them as specially fitted for occupation by this smaller and migratory member of this important family. The earlier published accounts of captures of Salmonidæ are very perplexing, since no two observers agree in their nomenclature, and one can never be quite sure which species is recorded. This difficulty need no longer exist, for the clear definitions laid down as the result of the masterly study of Salmonidæ by the late Dr. Day have simplified the matter for all future students. Yarrell (*British Fishes*, vol. ii., p. 39) notes the acquisition of one captured in the Thames, from the Shad fishermen who fish above Putney Bridge. Mr. E. A. Fitch records (*Essex Nat.*, vol. iii., p. 35) the capture of a specimen in the Blackwater, near Beeleigh.

Salmo fario, *Linn.* Brown Trout or Brook Trout.

Common as this fish is all over Britain, it is rare in Essex; and, excepting in the Lea, it only occurs in our streams as the result of artificial stocking. It was introduced in the Roman River, a feeder of the Colne, about twenty-five years ago, by the Rev. Mr. Marsh. In the Roding, it was introduced about 1881 by Mr. Rodwell, of High Laver, he having put into that river about 2,000 small Trout, which still appear to be doing well (*Essex Nat.*, vol. i., p. 149).

They have been taken in the small river at Great Chesterford (*Field*, July 9th, 1870, p. 25), and at Dagenham (*ib.*, June 16th, 1883). Numerous instances from the Lea are recorded, especially from Waltham Abbey and Sewardstone.

Mr. Miller Christy records (*Trans. Essex Field Club*, vol. i., p. 70) the capture of a specimen in the river Cann, at Chignal. This was diseased; but, generally speaking, it may be said that whenever the Brook Trout has been introduced it has flourished fairly well

CLASS PISCES.

Genus OSMERUS, *Artedi.*

Osmerus eperlanus, *Lacépède.* SMELT OR CUCUMBER SMELT.

Considerable numbers of this delicious fish are caught in all our rivers when they come up to spawn. Their peculiar cucumber-like scent is well known: hence the name. This distinguishes them at once from the worthless Atherine, a fish very similar in general appearance, but having no adipose fin.

Genus COREGONUS, *Artedi.*

Coregonus oxyrhynchus, *Rondel.* HOUTING.

Of this apparently-rare British fish, I saw, in 1886, several examples which had been captured in the smelt nets in the Colne. As it is frequently seen in boxes of Dutch Smelts, it is probably much more common in Holland than in this country. This is rather strange, when we consider the nearness of the Dutch coast.

Genus THYMALLUS. *Cuv.*

Thymallus vulgaris, *Nilss.* GRAYLING.

This is said to occur in the Cam, and may possibly be an inhabitant of that part of this river which rises in, and flows through, Essex. Dr. Day says (*Fishes of Great Britain*, vol. ii., p. 135) 1470 fry were placed in the Lea in 1863. I have not heard of the capture of any of them, but we must hope they are still doing well.

Family ESOCIDÆ, *Day.*

Genus ESOX, *Artedi, Cuv.*

Esox lucius, *Linn.* PIKE.

This, the gamest of our Essex fresh-water fish, is found in all parts of the county, both in rivers and ponds. It sometimes attains to a very large size in the Stour. In the Distillery ponds, at Colchester, it has been known to destroy brood after brood of young swans, a proof that here also it grows to a formidable size. Mr. Alfred Jardine captured one weighing 30lbs., on 22nd November, 1896, at Dagenham (*Badminton Mag.*, 1897, p. 628).

Family SCOMBRESOCIDÆ, *Day.*

Genus BELONE, *Cuv.*

Belone vulgaris, *Fleming.* GAR-FISH. Locally, GOREBILL.

Large numbers of this are taken at certain seasons round the coast, and they meet with a ready sale in London. Their peculiar green bones have not a tempting appearance, and I have no doubt prevent their being so generally appreciated as otherwise they would be.

Family SILURIDÆ, *Günth.*

Genus SILURUS, *Linn.*

Silurus glanis, *Linn.* SHEAT FISH, OR WELS.

The following note, from the pen of Dr. Günther (*Field*, September 8th, 1894, p. 411), relates the capture of one of this species in Essex waters :

"SIR,—Some thirty years ago, in the time of the Society of Acclimatisation, the proposal to introduce into Great Britain the Silurus or Wels of Central Europe was discussed in *The Field*. The suggestion did not meet with general approbation. The fish was described as an ugly-looking brute, a sort of fresh water shark, lurking among the bottom weeds, destroying a vast quantity of fish, and affording no sport When it became known, on the authority of an ancient chronicler, that the remains of a boy had once been taken from the stomach of a Wels, the report was conclusive as to the extreme undesirability of such an addition to the British fauna.

Perhaps it is just as well for the peace of mind of those who hold pessimistic views on the character of the Silurus, that they lived in ignorance of the fact that, all this time, individuals of this pædophagous monster were living a secluded life in more than one part of the country. Several gentlemen, who had personally become acquainted with this fish on the Continent, introduced a small number of young into more or less suitable private waters, where probably the majority found premature and nameless graves in the stomachs of Pike. At any rate, very little has been heard, and, as far as I know, nothing has been seen, of them. Therefore, the following instance of the capture of an adult specimen seems to be of particular interest. I am indebted to Mr. Nocton, of Langham Hall, Colchester, for my information.

A week or two ago, the gamekeepers of this gentleman, while engaged in Eel-fishing, caught a fish unknown to them, 4ft. 3in. long, and weighing over 30lbs., in the River Stour, at Flatford Mills, Suffolk. It was in perfect condition, but, probably owing to its having been rubbed by the net, it did not long survive.

Mr. Nocton suspected at once the real nature of the fish, and had it mounted by Mr. Gardner, of Oxford Street.

How did the fish get into the river Stour?

I am informed that Sir Joshua Rowley put, about twenty-nine years ago, young Siluri into a lake communicating with that river, and distant some six or seven miles from the place of capture.

There is, therefore, no doubt that this fish was a survivor of Sir Joshua's experiment, and that it grew in the intervening period to the size mentioned. Mr. Gardner tells me that it was a male fish.

I am unable to offer an opinion on the question whether this fish would have continued to grow if it had lived, or whether it had reached its maximum size long before capture. Male Siluri are generally smaller than females.

Austrian fishermen maintain that, in suitable localities, with an abundance of food and plenty of sea-room, the Wels can attain a weight of one and a half lbs. in the first year, and of three lbs. in the second; but, under less favourable conditions, the rate of growth is known to be much slower. An old ichthyologist (Balder) reports the case of a Wels which was one foot long when captured in the River Ill, in the year 1569. It was placed in a pond by itself, and had reached a length of five feet when it died in 1620.

In exceptionally favourable localities, as in the middle and lower courses of the Danube, Siluri are not rarely caught of 400 and 500 lbs., but there is no water in Great Britain in which this fish would ever be expected to grow to anything like that size. It would thrive in many an ornamental water, in which at present only coarse fish disport themselves; but neither from the angler's nor from the gastronome's point of view does it offer any particular inducement to introduce it. Living on such fish as are habitually feeding on the ground, it takes the bait only when sunk to the bottom, and at night-time, rather than during the day.

Young Wels frequently appear on the table where the fish is common; and, in the fish-market at Berlin, even larger specimens of 40 lbs. or 50 lbs. used to find a ready sale.

As I have not heard that the presence of Mr. Nocton's Silurus in the Stour has caused any great decrease in the population of that river, I am still of opinion that the introduction of this interesting species can be safely recommended to those who prefer the occasional excitement of an unusual capture to the monotonous landing of half-pound Roach; and I am very glad to find from this fortunate event that at least the practicability of acclimatising Silurus in Great Britain has been proved.

A. GÜNTHER."

Family CYPRINIDÆ, *Day.*

Genus CYPRINUS, *Artedi.*

Cyprinus carpio, *Linn.* CARP.

This is found in many of the sluggish streams. It also occurs in numerous ponds throughout Essex. I have seen an example, weighing 11½ lbs., which was taken in the Colne, at Earls Colne, on June 18th, 1889, and another weighing about 6 lbs., taken August 8th, 1890, in the flood-gate hole at East Mill, Colchester.

Genus CARASSIUS, *Nilss.*

Carassius vulgaris, *Nord.* CRUCIAN OR PRUSSIAN CARP.

This fish has been introduced into many of our ponds, but I do not know of its occurrence in any of our streams, except the Lea, where one was caught by Mr. Williams, near Tottenham (*Land and Water*, November 12th, 1887, p. 412).

Day says (*Fishes of Great Britain*, vol. ii., p. 166) it is very common about London.

Carassius auratus. *Bleeker.* GOLD FISH.

This is another fish which does not occur in our rivers. It is, however, naturalised in ponds throughout the county.

Genus BARBUS, *Cuv.*

Barbus vulgaris. *Fleming.* BARBEL.

This is probably another introduced species, as far as Essex is concerned. It occurs in the lake known as Dagenham Breach, and also, according to Day (*Fishes of Great Britain*, vol. ii., p. 171) and Yarrell (*British Fishes*, vol. i., p. 322), in the Lea. From that partially-Essex river, Barbel of nine and ten pounds weight are reported (*Field*, Jan. 5th, 1884, p. 18; *Land and Water*, Aug. 13th, 1887, p. 92).

Genus GOBIO, *Cuv.*

Gobio fluviatilis, *Flem.* GUDGEON.

A fish that occurs in numbers in all our rivers, but is mostly local. It abounds in the Lea, and is known in the Stort.

Genus LEUCISCUS, *Cuv.*

Leuciscus rutilus, *Fleming.* ROACH.

Found everywhere throughout the county, in streams and ponds.

Leuciscus cephalus, *Fleming.* CHUBB.

Until quite lately, this was another inhabitant of the Lea only among Essex rivers. Now it must be added to the list of Blackwater fish, as the Witham Angling Society has turned into that river 350 specimens (*Field*, Feb. 29th, 1896, p. 320). Lieut. Croft mentions it (*Trans. Herts. Nat. Hist. Soc.*, vol. ii., p. 13) in his list of fishes of the Lea, and it is frequently taken at Waltham.

Leuciscus vulgaris, *Fleming.* DACE.

A fish found in the Chelmer, the Stour, and, according to Lieut. Croft (*Trans. Herts. Nat. Hist. Soc.*, vol. ii., p. 13), in the Lea. Various catches are reported from Sewardstone (*Land and Water*, Sept. 24th, Nov. 5th, 1887, pp. 236 and 387). No doubt it occurs also in other streams. At times it gives good sport, as it takes the fly freely.

Leuciscus erythrophthalmus, *Fleming.* RUDD.

Considerable numbers of this fish occur in the Suffolk Stour. It is rare in the Colne, and common in the Lea (*Trans. Herts. Nat. Hist. Soc.*, vol. ii., p. 13.)

Mr. Pennell found (Day's *Fishes of Great Britain*, vol. ii., p. 184) a lemon or yellow-coloured variety of the Rudd in some ponds near Romford.

Leuciscus cœruleus, *Swainson.* BLUE ROACH OR AZURINE.

This fish is said to frequent an Essex stream, in support of which statement, I give the following quotation from Mr. Dorling's *Historical Guide to Walton-on-the-Naze and Neighbourhood*:

Inland fishing can be obtained for Roach, etc., some three miles to the west of Walton, in a little and apparently-insignificant river. But here as many as forty pounds of roach have been caught to a single rod in a day, many of the fish competing for size, colour, and flavour with those of the Thames or Lea. But, to the naturalist, as well as the angler, this stream will not fail of inducing a visit, as it contains that remarkably scarce fish in the British Islands, the Azurine or Blue Roach (*Leuciscus cœruleus*). Its habits are said to be much like those of the Chub, and especially is it highly retentive of life. It in shape resembles the Rudd; but, as regards colour, it is distinguished by having the upper part of the head, the back, and sides a slate blue, passing into silvery below, and both shining into a metallic lustre; whereas, in the Rudd, the lower part of the body is of a golden yellow. Also the fins of the Azurine are white, not, as in the Rudd, of a fine vermilion colour. (*Land and Water*, September 2nd, 1876, p. 153; *cf.* also *Field*, July 14th, 1877, p. 49.)

I have never been able to obtain a specimen. Although called a Blue Roach, this fish is really a Blue Rudd.

Leuciscus phoxinus, *Fleming.* MINNOW.

This fish is common in the Stour, the Colne, the Sandon brook (a feeder of the Chelmer), and probably also in all other Essex rivers. Its lively and active habits, and the readiness with which it may be tamed, make it a desirable occupant of the aquarium.

Genus TINCA, *Cuvier.*

Tinca vulgaris, *Cuvier.* TENCH.

A fish well known throughout the county as an inhabitant of ponds. Yarrell mentions (vol. i., p. 330) several localities for Essex Tench. It bears removal well, from its extraordinary tenacity of life, and this retentiveness has no doubt assisted in its general dispersion. There is every reason, however, for supposing that it is a true native of our larger rivers. I have had several specimens from the Stour and the Colne. Others are recorded from the Lea.

As regards the effect of the nature of the water in which it lives upon the suitability of the Tench for the table, Day (*Fishes of Great Britain*, vol. ii., p. 191) gives an instance of some taken from a fetid muddy pond at Mundon Hall, in this county, which were of an excellent flavour, while others procured from clear water at Leighs Priory smelt and tasted so rank that when dressed no one could eat them.

Genus ABRAMIS, *Cuvier.*

Abramis brama, *Fleming.* LAKE BREAM OR POMERANIAN BREAM, *var.*

This is found in considerable numbers in the Essex and Suffolk Stour, and in some other Essex rivers. The variety called Pomeranian Bream, is found, according to Day (*Fishes of Great Britain*, vol. ii., p. 195), on the authority of Yarrell, in Dagenham Breach, where it is still plentiful (*Land and Water*, August 4th, 1888, p. 134; *ibid.*, August 25th, p. 221; and *The Field*, June 18th, 1881, p. 835).

Abramis blicca, *Agass.* WHITE BREAM OR BREAM FLAT.

It occurs in the Essex and Suffolk Stour, also in the Lea (Lieut. Croft in *Trans. Herts. Nat. Hist. Soc.*, vol. ii., p. 12). It is not so common as the last species.

Genus ALBURNUS, *Heckel.*

Alburnus lucidus, *Heckel.* BLEAK.

Day says (*Fishes of Great Britain*, vol. ii., p. 200) this is found in the Lea. It is also in Lieutenant Croft's list. I do not think the Bleak is known in any other Essex river.

Genus NEMACHEILUS, *Van Hasselt.*

Nemacheilus barbatula, *Günther.* LOACH.

A fish generally distributed in all our streams, especially where the current is rapid and the bottom gravelly. It is very frequent at Lexden Springs. A lake at Elsenham Hall was said by Mr. A. Gilbey to abound with Loaches some years ago (*Field*, Feb. 7th, 1880, p. 153). Couch mentions the peculiar habits of this fish, which, at the approach of any striking change in wind or weather, becomes greatly excited, and often throws itself far out of the water.

Family CLUPEIDÆ, *Cuvier.*

Genus ENGRAULIS, *Cuvier.*

Engraulis encrasicholus, *Cuvier.* ANCHOVY.

Day (*Fishes of Great Britain*, vol. ii., p. 207) gives instances of the capture of this fish on the Essex Coast; and Yarrell (vol. i., p. 153) says: "It is reported to be at this time an inhabitant of the piece of water below Blackwall, called Dagenham Breach."

CLASS PISCES.

Genus CLUPEA, sp., *Artedi.*

Clupea harengus, *Linn.* HERRING.

This occurs commonly all round the coast. It was formerly taken in the Estuary of the Blackwater in sufficient numbers to make it worth while to fish for it with drift-nets after the manner followed (although on a much larger scale) in the North Sea. Of late years, considerable numbers of Whitebait, satisfactorily proved to be Herring fry, have been taken in the Crouch and the Blackwater, and dispatched to London.

Clupea pilchardus, *Artedi.* PILCHARD.

Is occasionally taken on the coast. Dale gives it in his list of Harwich fish, and says (*Hist. of Harwich*, p. 432): "It is rare, but is occasionally brought among Herrings to market." Day says (*Fishes of Great Brit.*, vol. ii., p. 230) Yarrell obtained one in May, 1838, from the mouth of the Thames.

Clupea sprattus, *Linn.* SPRAT.

Enormous numbers of this fish are taken off our coast by the stow-boat fishermen. A good Sprat season is the great harvest of the Wyvenhoe and Brightlingsea men, and plenty of Sprats means to these places a prosperous winter. An immense proportion of the fish taken is used for manure, but very large quantities are sold in a fresh state for food. The Aldeburgh plan of fishing with drift-nets is not so destructive to the immature, not only of sprats but of all other kinds of fish, as is our stow-boat fishing.

Clupea alosa, *Linn.* ALLIS SHAD.

I have never taken this fish. Day (*Fishes of Great Britain*, vol. ii., p. 236), quoting Yarrell, mentions one being taken in the Thames above Putney in 1831. Jenyns describes it (*Brit. Vert. Animals*, p. 438) as "occasionally, though rarely, taken in the Thames."

Clupea finta, *Cuvier.* TWAIT SHAD.

This is occasionally common in the Colne. Many were taken by the Smelt fishermen in August, 1886, and one was caught on November 29th, 1886, at East Bridge, Colchester, which point is the tidal limit. Jenyns says (*Brit. Vert. Animals*, p. 438) it is very abundant in the Thames.

Family MURÆNIDÆ, *Müller.*

Genus ANGUILLA, *Willugh.*

Anguilla vulgaris, *Turton.* EEL.

This fish is found in all our Essex rivers. Dr. Day, in his work on *The Fishes of Great Britain*, so frequently quoted in these pages, has conclusively shown that the numerous forms of Sharp-nosed, Broad-nosed, and Snig-Eels are only varieties of this species, which is the only Eel inhabiting fresh waters—at least, for a part of its existence. It seems, at first sight, a mystery how isolated pieces of water can have become stocked by this fish; but, if we remember that the "Elvers," or young Eels, are always moving up stream, following even the smallest trickling to its source, the explanation becomes comparatively easy, since there are very few ponds that do not at some time or other overflow, and the slightest stream is sufficient for these tiny wanderers.

The well-known capability also of the larger Eels to wriggle over considerable distances of land through wet grass or other herbage must not be forgotten, although this, as far as my experience goes, more frequently occurs when the pools are drying up.

One instance, I remember, where the water in a large ditch, forming part of the fence of a garden, had nearly evaporated. By the help of a little dog, I caught in the grass several large Eels, which were evidently on their way to the brook, some seventy or a hundred yards distant. These, probably, had at some time or other passed from the brook to the garden fence.

Some of the most remarkable ichthyological observations of recent years are those of Prof. Grassi on the development of some species of Murænidæ, and particularly of the Common Eel, which have been communicated to the Royal Society (*Proc. Roy. Soc.*, vol. lx., pp. 260-271), and are summarised in the *Essex Nat.* (vol. ix., p. 261). The larva of the Eel is the little fish called by Pennant *Leptocephalus brevirostris*, which lives in abysmal waters, and is very rarely found floating on the surface. The form described by Yarrell as *L. morrisii* has been proved by Grassi to be the larva of the Conger Eel. The whole subject is one of the greatest interest to naturalists, and is an admirable example of the necessity and value of careful and long-continued observation of even the commonest species.

I consider Eels the very finest and most delicious of our fresh-water fish, especially after they have left our rivers and taken up their residence in salt water. This statement may partake somewhat of the character of an Irish bull; but the meaning I wish to convey is that an Eel from salt water is free from muddy flavour, and is in every way superior to the same fish during its residence in fresh water.

Genus CONGER, *Cuvier.*

Conger vulgaris, *Cuvier.* CONGER.

Our sandy coast is not suited to the habits of this fish. An occasional one is, however, taken. One of 40 lbs. weight is recorded (*Essex Standard*, Jan. 31st, 1885) as having been picked up on the beach at Clacton-on-Sea. Day, quoting Donovan, says (*Fishes of Great Britain*, vol. ii., p. 253) one of 130 lbs. weight was captured at the Nore. In the *Zoologist*

(1869, p. 1520) a record is given of several which were caught in the Thames as high as Woolwich.

Mr. E. A. Fitch says (*Essex Nat.*, vol. i., p. 218) one weighing 35 lbs. was captured on October 26th, 1887, at Beeleigh, and the capture of two off Clacton is recorded (*Essex Nat.*, vol. ii., p. 6). About the middle of February, 1894, a police-constable named Harrington shot one, which weighed nearly 30 lbs., from the sea-wall of Foulness Island (*Essex County Chronicle*, February 23rd, 1894).

Order IV. LOPHOBRANCHII, *Cuvier.*

Family SYNGNATHIDÆ, *Linn.*

Genus SYPHONOSTOMA, *Kaup.*

Syphonostoma typhle, *Kaup.* BROAD-NOSED PIPE-FISH.

This is very common on our *Zostera*-covered shores. It is taken very frequently by the shrimp- and eel-trawlers.

Genus SYNGNATHUS, *Artedi.*

Syngnathus acus, *Linn.* GREATER PIPE-FISH.

This, the commonest of the family, is found on all parts of our coast, but more frequently on the beds of the sea-wrack (*Zostera marina*), to the tufts of which it clings. It is, therefore, constantly captured in the eel-trawls, and forms one out of the mass of living creatures brought by these nets to the surface.

Genus NEROPHIA, *Rophinesque.*

Nerophis æquoreus, *Kaup.* OCEAN, OR SNAKE, PIPE-FISH.

Although not so common a fish as the preceding, this is found in some quantity in the eel-trawls. Day (*Fishes of*

Great Britain, vol. ii., p. 262) gives the estuary of the Thames as their habitat.

Nerophis ophidion, *Linn.* STRAIGHT-NOSED, OR SNAKE, PIPE-FISH.

Buckland says (*Fam. Hist. of Brit. Fishes*, p. 197) this is often taken with Whitebait in the mouth of the Thames.

Genus HIPPOCAMPUS.

Hippocampus antiquorum. SEA-HORSE

Dr. Bree records (*Field*, December 1st, 1866, p. 420) the capture of two specimens of this rare fish at Brightlingsea. I have never seen it myself.

Order V. PLECTOGNATI.

Family GYMNODONTES.

Genus ORTHAGORISCUS, *Bl., Schu.*

Orthagoriscus mola, *Bl., Schu.* SHORT SUN-FISH.

Lindsey says (*A Season at Harwich*, p. 102) this occurs but occasionally on the Essex coast. Mr. E. A. Fitch informs me that he saw a very large one which had been taken in the River Crouch, at Battles Bridge, on October 21st, 1874. It was a female, measuring 4ft. 6in. in length, and was carried to Rayleigh, Rochford, and Chelmsford for exhibition (*Land and Water*, October 31st, 1874, p. 340). Some years later, Mr. Fitch rowed after another in the same river at Burnham. It moved very slowly, and allowed his boat to approach quite closely to its fin before it disappeared.

Orthagoriscus truncatus, *Fleming.* OBLONG SUN-FISH.

Lindsey says (*A Season at Harwich*, pt. ii., p. 101) the Oblong Sun-fish is rare.

Sub-Class CHONDROPTERYGII.

Order GANOIDEI.

Sub-Order CHONDROSTEI.

Family ACIPENSERIDÆ.

Genus ACIPENSER, *Artedi.*

Acipenser sturio, *Linn.* STURGEON.

This is occasionally taken in all our rivers, but it is rare. When captured in the Thames, within the jurisdiction of the City of London, it is usually judged a proper present for the Lord Mayor's table.

The capture of a Sturgeon, weighing 131 lbs., in the Blackwater River, near Beeleigh Mills, Maldon, is noted as a remarkable circumstance (Donovan, *British Fishes*, vol. iii., plate lxv.) Dr. Bree, referring to a small specimen found at Thorpe-le-Soken in 1864, raises the question whether the fish breeds here or migrates from the Danube (*Field*, Aug. 6th, 1864, p. 105).

Mr. Fitch records the capture of two very large Sturgeons in the Blackwater on May 9th, 1886, and May 15th, 1890. The latter, which weighed 212 lbs. and measured seven feet eleven inches in length, was exhibited at Chelmsford, and eventually went to Sweeting's, in Cheapside (*Essex Nat.*, vol. iv., p. 120). Others are mentioned (*Zool.*, 1879, p. 383, and 1883, p. 341). About the middle of August, 1891, those engaged on the smack "Emmie," whilst dredging near Cliff Reach, Burnham, saw a fine Sturgeon disporting itself (*Essex County Chronicle*, August 21st, 1891).

CLASS PISCES.

Order ELASMOBRANCHII.

Sub-Order PLAGIOSTOMATA.

Family CARCHARIIDÆ.

Genus GALEUS, *Cuvier.*

Galeus vulgaris, *Fleming.* TOPER OR SWEET WILLIAM.

This Shark is, in my experience, but rarely taken on our coast. One, caught by trawling in the Wallet, on October 23rd, 1886, was thought by its captor of sufficient rarity to make it worth exhibiting in the town of Colchester. The capture of another below Brightlingsea, in a shrimp-trawl, is noted (*Essex Nat.*, vol. ii., p. 137). This was a female, measuring four feet nine inches long. Another, also a female, and five feet six inches in length, was captured near Clacton pier, on November 1st, 1888 (*Essex Nat.*, vol. ii., p. 256).

Family LAMNIDÆ.

Genus LAMNA, *Cuvier.*

Lamna cornubica, *Cuvier.* PORBEAGLE OR BEAUMARIS SHARK.

A fine specimen of this Shark was captured in a herring-net on the Essex coast in 1874. It measured eight feet two inches in length, and weighed from sixteen to twenty stone (*Land and Water*, Oct. 10th, 1874, p. 285). Another was taken off Harwich, by a lugger fishing for mackerel, and was exhibited at Colchester and Ipswich. It measured between nine and ten feet in length (*Essex Nat.*, vol. vi., p. 154). Soon after, on December 8th, 1892, a third was secured by Mr. Henry Gentry, of St. Osyth, in the estuary of the Colne. This one, a male, was seven feet ten inches long (*ibid.*, p. 207).

Family SPINACIDÆ.

Genus ACANTHIAS, *Risso.*

Acanthias vulgaris, *Risso.* PICKED DOG-FISH.

This voracious fish is always much too common, and in some years more especially so. It neither leaves a bait on the hook nor hardly an untouched fish in the fishermen's net. In handling it, great care should be taken to avoid its dangerous spines.

Family RHINIDÆ.

Genus RHINA, *Klein.*

Rhina squatina, *Belon.* MONK-FISH.

This frequents the entire Essex coast. It is usually caught in nets. Though occasionally eaten by fishermen, it is, according to my taste, far too rank in flavour for a more delicate palate.

Family RAIIDÆ.

Genus RAIA, *Artedi.*

Raia batis, *Linn.* SKATE.

This creature is not rare, especially on the sandy parts of the coast. The young of this and other Skates are called "maids," and are much esteemed for the table. The full-sized fish is also very good when properly dressed.

Raia alba, *Lacépède.* SHARP-NOSED SKATE.

Lindsey says (*A Season at Harwich*, pt. ii., p. 98) the French, who are great consumers of Skate, this species being their favourite, would come to Harwich for this fish, if greater encouragement were given to the fishery. As it is, he remarks, they go to Plymouth, and purchase large quantities during the season of Lent.

Raia clavata, *Linn.* THORNBACK.

This very common fish is found in more shallow water than any other species of Skate. It is usually esteemed the best of the family for the table.

Family TRYGONIDÆ.

Genus TRYGON, *Adanson.*

Trygon pastinaca, *Cuv.* STING RAY, OR FIRE FLAIRE. Locally, "FIERCE CLAW."

I have occasionally taken this fish when trawling in the estuary of the Blackwater.

It is more frequently taken in the channels between the sands, if I may judge from the number of needles, made from the spines upon its tail, that I have seen in the possession of fishermen, who use them for stringing flat-fish together through the gills, for the convenience of carriage.

Dr. Day says (*Zool.*, 1881, p. 386) they are by no means uncommon off the mouth of the Thames.

Sub-Class CYCLOSTOMATA.

Family PETROMYZONTIDÆ.

Genus PETROMYZON, *Artedi.*

Petromyzon marinus, *Linn.* SEA LAMPREY.

This is captured occasionally in the mouths of all our rivers. I once saw a very fine one that had been taken in the Colne, and I have recorded (*Essex Nat.*, vol. v., p. 134) the capture of another in the same river. I never knew anyone here who tried to eat his capture, most inhabitants of Essex considering them too loathsome in appearance. This, however, is not the case in districts where they are more frequent. Yarrell says (*British Fishes*, vol. ii., p. 451) one was taken in June, 1834, and another in the same month in 1835, as high up the Thames as Sunbury Weir. Day (*Fishes of Great Britain*, vol. ii., p. 358) gives for their habitat the mouth of the Thames.

Petromyzon fluviatilis, *Linn.*   LAMPERN OR SILVER LAMPREY.

Dr. Day (*Fishes of Great Britain*, vol. ii., p. 360, 361) says they are captured in the Thames, where they abound in quantities. Jenyns says (*Brit. Vert. Animals*, p. 521) they are common in many of our rivers, especially in the Thames. The same remark is made by Yarrell (*British Fishes*, vol. ii., p. 454). Lieutenant Croft mentions them (*Trans. Herts. Nat. Hist. Soc.*, vol. ii., p. 15) as found in the Lea.

They probably occur in all Essex rivers, but, as they are never fished for, I have not had the opportunity of seeing an example.

Petromyzon branchialis, *Linn.*   PLANER'S LAMPREY.
MUD LAMPREY (immature form).

According to my experience, this Lamprey is rare in Essex streams. I have, however, taken specimens at Lexden, in the Colne, and also in the Roman river, a feeder of the Colne.

ADDENDA.

Genus VESPERTILIO (p. 35).

[*Add on p.* 37 : **Vespertilio murinus.** MOUSE-COLOURED BAT.

Mr. E. Newman says (*Field*, March 14th, 1874, p. 263) that, many years ago, Doubleday shot at Epping a Bat which he believed to be of this species, after watching it on the wing for some time ; but, unfortunately, the specimen, though searched for diligently at the place where it fell after being shot, could not be found.]

Genus VIPERA (p. 85).

Vipera berus. ADDER OR VIPER.

Add on p. 86 : A careful description of a male specimen taken in Epping Forest was communicated by Mr. G. A. Boulenger to *The Essex Naturalist* in 1896 (vol. ix., p. 81).

Genus RANA (p. 86).

Rana temporaria. COMMON FROG.

Add on p. 86 : Mr. William Cole states (*in litt.*) that, so far as he has been able to discover, it does not occur on Mersea Island.

Genus ORCYNUS (p. 92).

Orcynus thynnus. COMMON TUNNY.

Add on p. 92 : In *The Essex County Chronicle* of Oct. 26th, 1897, Mr. H. L. Matthams records the finding of a big fish on the shore at Foulness on the 20th of that month. It was perfectly fresh, measured nine feet in length, and weighed about five or six hundredweight. From his description, it could be no other than the Short-finned Tunny. He also has stated in a letter to me that, about forty years ago, a large fish of probably the same species came ashore in the same island (see also *Zoologist*, 1897, p. 579).

Genus MOTELLA (p. 100).

Add on p. 100: **Motella tricirrata,** *Gmel.* THREE-BEARDED ROCK LING.

This has been taken in the River Blackwater by Mr. E. A. Fitch.

Genus SOLEA (p. 102).

Add on p. 102: **Solea lascaris,** *Risso.* LEMON, OR SAND, SOLE.

An Essex specimen of this fish is preserved at the Marine Biological Station of the Essex County Council at Brightlingsea.

Add on p. 102: **Solea lutea,** *Bonap.* LITTLE, OR RED, SOLE.

A specimen of this fish—the Solonette of Couch (*Fishes of British Isles*, vol. iii., p. 207)—exists also at the Marine Biological Station at Brightlingsea.

APPENDIX.

A LIST OF THE PRINCIPAL AUTHORITIES REFERRED TO IN THIS WORK.

Chronicles, by RALPH HOLINSHED, London, 1587, folio.
Annals, by JOHN STOW, London, 1605, folio.
Britannia Baconia, by JOSHUA CHILDREY, London, 1660, 8vo.
History of the Worthies of England, by THOMAS FULLER, London, 1662, folio.
Theatre of the Empire of Great Britain, by JOHN SPEED, London, 1676, folio.
The History and Antiquities of Harwich and Dovercourt, by SILAS TAYLOR, edited, with appendix, by SAMUEL DALE, 2nd edition, London, 1732, 4to.
History of Waltham Abbey, by JOHN FARMER, London, 1735, 8vo.
Survey of the Country Fifteen Miles round London, by JOHN CARY, London, 1796, 8vo. (MS. Note by CARY in Mr. B. G. Cole's copy).
Rural Sports, by WILLIAM BARKER DANIEL, two volumes, London, 1801-2, 4to; (2nd edition, three volumes, 1812, to which reference is made); Supplement, 1813.
Natural History of British Fishes, by EDWARD DONOVAN, five volumes, London, 1802-8, 8vo.
The Sportsman's Vocal Cabinet, edited by CHARLES ARMIGER, London, 1830, 8vo.
Proceedings of the Zoological Society of London, 1833 to date.
Sketch of the Natural History of Great Yarmouth, by CHARLES JOHN and JAMES PAGET, Yarmouth, 1834, 8vo.

A Manual of British Vertebrate Animals, by the REV. LEONARD JENYNS, M.A., Cambridge, 1835, 8vo.

"An Account of Fishes found in Norfolk, and on the Coast," by SIR THOMAS BROWNE. *Works*, London and Norwich, 1835, vol iv., p. 325.

A History of British Fishes, by WILLIAM YARRELL, two volumes, London, 1836-60, 8vo.

Speculi Britanniæ pars: An Historical and Chorographical Description of the County of Essex, by JOHN NORDEN (1594), edited by Sir H. ELLIS for the Camden Society, London, 1840, 4to.

Annals and Magazine of Natural History, 1841, etc., 8vo.

The Zoologist, a popular Miscellany of Natural History, edited by EDWARD NEWMAN, J. E. HARTING, and W. L. DISTANT, London, 1843 to date. 8vo.

Observations on the Fauna of Norfolk, by the REV. RICHARD LUBBOCK, Norwich, 1845, 8vo.

Autobiography of Sir John Bramston, edited by Lord BRAYBROOKE for the Camden Society, London, 1845, 4to.

A History of British Reptiles, by THOMAS BELL, 2nd edition, London, 1849, 8vo.

A Season at Harwich, by W. H. LINDSEY, London, 1851, 8vo.

The Field, or Country Gentleman's Newspaper, 1853 to date.

Land and Water, 1866 to date.

History of Rochford Hundred, by PHILIP BENTON, Rochford, 1867, etc., 8vo (incomplete).

English Deer Parks, by EVELYN P. SHIRLEY, London, 1867, 8vo.

Familiar History of British Fishes, by FRANK BUCKLAND, London, 1873, 8vo.

"A List of the Fishes known to occur in the Norfolk Waters," by JOHN LOWE, M.D., in *Transactions of the Norfolk and Norwich Naturalists' Society*, Norwich, 1873-4, 8vo, p. 25.

A History of British Quadrupeds, including the Cetacea, by THOMAS BELL, F.R.S., 2nd edition, London, 1874, 8vo.

The Natural History of Selborne, by GILBERT WHITE, edited by J. E. HARTING, London, 1875, 8vo.

Monograph of Asiatic Chiroptera, and a Catalogue of the Bats in the Indian Museum, Calcutta, by G. E. DOBSON, London, 1876.

Catalogue of the Chiroptera in the Collection of the British Museum, by G. E. DOBSON, F.R.S., London, 1878, 8vo.

The Fishes of Great Britain and Ireland, by FRANCIS DAY, two volumes, London, 1880–84, 8vo.

Transactions of the Essex Field Club, Buckhurst Hill, 1880–1886, 8vo.

Journal of the Proceedings of the Essex Field Club, Buckhurst Hill, 1880–86, 8vo.

Seals and Whales of the British Seas, by THOMAS SOUTHWELL, F.Z.S., Norwich, 1881.

A Handbook of the Vertebrate Fauna of Yorkshire, by W. EAGLE CLARKE, and W. DENISON ROEBUCK, London, 1881.

The Essex Naturalist, Buckhurst Hill, 1887 to date.

Articles on "Izaak Walton and the River Lea," and "The River Lea below Hertford," by Lieut. R. B. CROFT, R.N., in *Transactions of the Hertfordshire Natural History Society and Field Club,* Watford, Hertford, vol. ii., pp. 9, 243.

The Forest of Essex: Its History, Laws, Administration, and Ancient Customs, and the Wild Deer which lived in it, by W. R. FISHER, London, 1887, 4to.

The Birds of Essex: A Contribution to the Natural History of the County, by MILLER CHRISTY, London, etc., 1890, 8vo.

A Descriptive List of the Deer Parks and Paddocks of England, by JOSEPH WHITAKER, London, 1892, 8vo.

The Essex Foxhounds, with Notes upon Hunting in Essex, by R. F. BALL and TRESHAM GILBEY, London, 1896, 4to.

Epping Forest, by EDWARD NORTH BUXTON, Verderer, 4th edition, London, 1897, obl. 8vo.

APPENDIX B.

LIST OF SUBSCRIBERS.

(In the following list, the names marked by an asterisk (*) are those of members of the Essex Field Club, who are, by arrangement, entitled to subscribe for a single copy at a reduced price.)

Abbott, G., M.R.C.S., 33, Upper Grosvenor Road, Tunbridge Wells.
Acland, Rev. C. L., M.A., F.S.A., All Saints, Cambridge.
* Adams, H. J., F.E.S., Roseneath, Enfield.
Aflalo, F. G., The Club, Bournemouth.
Atkinson, Rev. Canon J. C., Danby Vicarage, Yarm, Yorks.
* Avery, J., C.A., 63, Windsor Road, Forest Gate, E.
Ball, Rev. F. J., Ardleigh Vicarage, Colchester.
Barclay, F. H., 4, Second Avenue, West Brighton.
Barclay, Col. Hanbury, Tingrith Manor, Woburn, Beds.
Barrington, R. M., F.L.S., Fassaroe, Bray, Co. Wicklow.
* Bartlett, H. S., Severndroog, Shooter's Hill, Kent.
Bawtree, E. W., M.D., St. Mary's, Colchester.
Beckett, Rev. W. H., Stebbing Manse, Chelmsford.
Bentall, E. H., The Towers, Maldon.
* Bewers, W., Dorincourt, Chelmsford.
* Bird, James, Church Hill, Walthamstow, Essex.
* Bird, T., C.C., Canons, Romford.
Blathwayt, Frank L., Saltaire, Weston-super-Mare, Somerset.
Borrer, Wm., Cowfold, Horsham, Sussex.
* Boulger, Prof. G. S., F.L.S., F.G.S., Permanent V.P., 34, Argyll Mansions, Addison Bridge, W.
* Brown, George, M.D., Headgate, Colchester.
Burke, Lt.-Col., J.P., Auberies, Sudbury, Suffolk.
Cambridge, Rev. O. P., Bloxworth Rectory, Wareham.
Catchpool, Edward, Feering Bury, Kelvedon, Essex.
* Chancellor, F., J.P., F.R.I.B.A., Permanent V. P., Chelmsford.
* Chattaway, W., Apothecaries' Hall, London.
* Christy, Reginald W., Little Boyton Hall, Roxwell, Chelmsford.
* Christy, Miller, F.L.S., Pryors, Broomfield, Chelmsford.
* Clark, J. A., M.P.S., L.D.S., &c., 57, Weston Park, Crouch End, N.
Cobbold, A. Townshend, Fermoyle, Ipswich. (2 *copies*.)
* Coburn, H. I., Hon. Solicitor, 16, Maxilla Gardens, N. Kensington, W.

* Cole, William, F.E.S., F.L.S., Hon. Secretary, 7, Knighton Villas, Buckhurst Hill, Essex.
Coleman, The Misses, 6, St. Mary's Terrace, Lexden Road, Colchester.
* Coles, F., 53, Brooke Road, Stoke Newington, N.
* Colvin, R. B., J.P., &c., Monkhams, Waltham Abbey.
Connor, Colonel, Elmhurst, Festing Road, Southsea.
Cordeaux, John, J.P., F.R.G.S., M.B.O.U., Great Cotes House, R.S.O., Lincolnshire.
Cotgreave, A., The Public Libraries, West Ham. (*2 copies.*)
Cotter, Rev. J. Rogerson, Magdalen Rectory, Colchester.
* Crouch, Walter, F.Z.S., V.P., Grafton House, Wellesley Road, Wanstead. (3 *copies.*)
Crowfoot, W. M., Beccles, Suffolk.
* Crump, Miss Rhoda, St. Aubyn's, Woodford Green.
* Cunnington, A., Braintree.
Deedes, Rev. Canon Cecil, 2, Clifton Terrace, Brighton.
De Horne, Thomas, 5, Stanhope Terrace, Hyde Park, W.
* Dewick, The Rev. E. S., M.A., F.S.A., F.G.S., 26, Oxford Square, Hyde Park, W.
Downes, Arthur, M.D., 46, Gordon Square, W.C.
Duff, The Right Hon. Sir M. E. Grant, P.C., Lexden Park, Colchester.
* Duffield, F. H., St. Oswald's House, Shortlands, Kent.
Dunnage, Alfred, Ivy Lodge, Dedham, Colchester.
* Egerton-Green, C. E., East Hill House, Colchester.
* Elliott, F. W., Mark Ash, Palmerston Road, Buckhurst Hill.
* Fitch, E. A., F.L.S., F.E.S., C.C., Permanent V.P., Brick House, Maldon.
Fletcher, W. H. B., Fairlawn House, Worthing, Sussex.
* Float, J. C., Fullbridge, Maldon.
Fortune, Riley, Alston House, Harrogate.
* Fountain, F., 44, Crooms Hill, Greenwich Park, S.E.
* Furbank, A. J., 1, Cardigan Road, Richmond Hill, Surrey.
Galpin, Rev. Francis W., Hatfield Vicarage, Harlow, Essex.
Gates, Arthur F., 21, Dudley Road, Ilford, Essex.
* George, W., 19, Church Crescent, South Hackney, N.E.
* Gepp, The Rev. E., M.A., School House, Felstead.
Gilbert, T., J.P., West Mersea, Colchester.
Gilbey, Sir Walter, Bart., Elsenham Hall, Essex, and Cambridge House, Regent's Park, N.W.
Gilmour, Matthew A. B., F.Z.S., Saffronhall House, Windmill Road, Hamilton, N.B.
* Gimson, W. G., M.D., &c., Witham, Essex.
Gordon & Gotch, 15, St. Bride Street, E.C.
* Gould, I. Chalkley, Trapp's Hill House, Loughton.
* Gray, Mrs. C. Harrison, Laurel Grove, Chelmsford.
* Greatheed, F., Corringham, Stanford-le-Hope, S.O., Essex.
Green, E. R., 26, Throgmorton Street, E.C.
Green, Horace G. Egerton, J.P., King's Ford, Colchester.
Gurney, J. H., F.Z.S., &c., Keswick Hall, Norwich.
Haigh, G. H. Caton, Grainsby Hall, Great Grimsby, Lincolnshire.
* Hall, Samuel, F.C.S., 19, Aberdeen Park, Highbury, N.

LIST OF SUBSCRIBERS.

Hall, Rev. Edmund, Myland Rectory, Colchester.
* Harrison, A. W., Blandford House, Braintree.
* Harrison, A., 72, Windsor Road, Forest Gate.
* Harting, J. E., F.L.S., &c., Librarian to Linneau Society, Burlington House, Piccadilly, W.
Harvie-Brown, John A., Dunipace House, Larbert, N.B.
* Hawkins, C. E. W., Old House, Great Horkesley, Colchester.
Hawkins, C. H., J.P., Maitlands, Colchester.
Hills, Harris, J.P., Berewyk Hall, near Halstead, Essex.
* Holmes, E. M., F.L.S., &c., Ruthven, Sevenoaks, Kent.
* Holmes, T. V., F.G.S., M.A.I., 26, Crooms Hill, Greenwich Park, S.E.
* Hope, G. P., Havering Grange, near Romford.
Horton, J. H., J.P., Mascalls, Brentwood.
Horwood, Rev. E. R., J.P., The Vicarage, Maldon, Essex.
* Howard, David Lloyd, J.P., F.I.C., F.C.S., President, Devon House, Buckhurst Hill.
* Howard, W. Dillworth, 11, Cornwall Terrace, Regent's Park, N.W.
* Howard, Eliot, J.P., Ardmore, Buckhurst Hill.
* Hughes, F., 17, Fairfield Road, Chelmsford.
* Hunt, Reuben, J.P., Tillwicks, Earls Colne, Essex.
* Hurnard, S. F., J.P., Lexden, Colchester. (3 *copies*.)
Hurrell, Wm. A., Southminster, Essex.
* Johnson, S. H., M.I.M.E., Warren Hill, Loughton. (2 *copies*.)
* Kemble, T., J.P., Runwell Hall, Wickford.
* Kimbell, H. I., 381, Hornsey Road, N.
* Knight, J. W., F.G.S., Bushwood, Wanstead.
Kuypers, Rev. Charles, St. Edmund's College, Ware.
La Barte, Rev. W. W., M.A., F.A.I., 9, Creffield Road, Colchester.
* Landon, F., The Red House, Brentwood.
Laver, Arthur H., M.D., 1, Rutland Park, Sheffield.
* Leach, H. R., Hazeldean, Pinner, Middlesex.
Lloyd, Llewelyn, Tendring, Colchester.
* Lockyer, Alfred, Mulberrys, Nazing, Waltham Cross.
* Lodge, Robert, Laurie Lodge, Meads, Eastbourne.
Loyd, George, 64, North Hill, Colchester.
Macpherson, A. Holte, 51, Gloucester Terrace, Hyde Park, W.
* Maitland, Rev. J. Whitaker, M.A., Loughton Hall, Loughton, Essex.
* Mason, Hugh H., Abbey Lodge, Barking, Essex.
* McConnell, P., B.Sc., Ongar Park Hall, Ongar, Essex.
* Meldola, Prof. R., F.R.S., F.R.A.S., F.C.S., Permanent V.P., 6, Brunswick Square, W.C.
Midgley, Arthur, Larchmont, Saffron Walden.
Moor, Captain E. C., The Rosery, Great Bealings, Woodbridge.
* Morris, H. G., 269, Lewisham High Road, St. John's, S.E.
* Mothersole, H., Park Avenue, Chelmsford.
Munsey, W. J., 5, High Street, Colchester.
Murton, Rev. George, 51, Lexden Road, Colchester, Essex.
* Newling, Benjamin, Woodleigh, South Woodford, Essex.
* Newton, E. T., F.R.S., F.G.S., Geological Museum, 28, Jermyn Street, Piccadilly, W.

* Nichols, W. B., East Lodge, Mistley, Manningtree. (3 *copies*.)
  Ogilvie, F. Menteith, 5, Evelyn Mansions, Carlisle Place, S.W.
* Oldham, Charles, 2, Warwick Villas, Chelmsford Road, South Woodford.
  Oldham, Charles, 58, Fountain Street, Manchester.
  Osborne, Arthur T., Altnacealgach, Colchester.
* Parker, C. W., J.P., Hatfield Priory, Witham.
* Paterson, W. M., South Lawn, Bishops Stortford, Herts.
  Patterson, R. Lloyd, Holywood, Co. Down.
  Paxman, James, J.P., Stisted Hall, Braintree.
  Piggot, Horatio, 20, Broadwater Down, Tunbridge Wells.
  Pointing, W. J., 58, North Hill, Colchester.
* Potter, J. W., Trinity Street, Colchester.
* Poulton, Prof. E. B., M.A., F.R.S., F.L.S., F.E.S., &c., Wykeham House, Banbury Road, Oxford.
* Power, H., M.B., F.L S., 37A, Great Cumberland Place, W.
* Prance, The Rev. L. N., M A., F.S.A., Stapleford Tawney, Romford.
* Pritchett, G. E., F.S.A., Oak Hall, Bishops Stortford.
  Reid, James, 12, Lower Bridge Street, Canterbury.
  Rivers, H. Somers, Sawbridgeworth, Herts.
* Roberts, F. G. Adair, Lion House, Amhurst Park, Stamford Hill, N.
  Rope, Geo. Thos., Blaxhall, Wickham Market, Suffolk.
* Rose, A. B., St. Helens, 56, Woodberry Down, N.
  Rothschild, The Hon. Walter, F.L.S., New Court, St. Swithin's Lane, E.C.
* Round, James, M.P., Birch Hall, near Colchester.
* Royle, Mrs. A., 329, Upton Lane, Forest Gate, E.
* Rudler, F. W., F.G.S., 28, Jermyn Street, S.W
* Russell, Rev. A. F., M.A., Chairman, Epping Forest Museum Committee, The Chantry, Chingford, Essex.
  Salter, J. H., J.P., M.D., &c., Tolleshunt D'Arcy. Witham, Essex.
* Sauzé, H. A., 4, Mount Villas, Sydenham Hill Road, S.E.
  Savill, Lt -Col., J.P., &c., Boleyns, Braintree.
* Schwartz, Cecil, Ivy House, Woodford Green, Essex.
  Scott, T., 158, Maldon Road, Colchester.
* Shenstone, J. C., 13, High Street, Colchester.
* Short, W. H., Sunnyside, Colchester.
  Silver, S. W., Letcomb Manor, Wantage.
  Simonds, Tom, Wavery, Hainhault Road, Leytonstone.
* Smith, Fred. J., 4, Christopher Street, Finsbury Square, E.C.
* Smithers, H. W., Baddow Court, Great Baddow.
* Sorby, H. C., LL.D., F.R.S., Broomfield, Sheffield.
  Southwell, Thomas, 10, The Crescent, Norwich.
  Spalding, Frederick, The Museum, Colchester.
  Sparrow, Miss E. B., Rookwoods, Sible Hedingham, Essex.
* Squier, S. W., J.P., The Rookery, Stanford-le-Hope.
* Stable, D. Wintringham, LL.B., Holly Lodge, Wanstead, Essex.
  Standen, R. S., F.L.S., Thorpe Hall, Colchester.
* Steele, A. R., Northbrooks, Harlow, Essex.
  Symmons, Robert F., 9, Lexden Road, Colchester.
* Taylor, Shephard T., M.B., Mount Echo, Chingford.
  Taylor, Thomas, Bocking, Braintree, Essex.

LIST OF SUBSCRIBERS.

Thackeray, Lt.-Col., J.P., Chappel, Halstead, Essex.
* Thresh, J. C., D.Sc., M.D., D.P.H., &c., The Limes, Chelmsford.
* Tower, C. J. H., J.P., D.L., Weald Hall, Brentwood.
Tremlett, J. D., J.P., Dalethorpe, Dedham, Essex.
* Tuke, Wm. Murray, Saffron Walden.
Unwin, George, 27, Pilgrim Street, E.C.
* Usborne, Mrs., Writtle, Chelmsford.
* Vaughan, G. E., Chingford, Essex, and 57, Chancery Lane, W.C.
* Walker, Henry, F.G.S., 150, Kensington Park Road, W.
* Waller, W. C., M.A., F.S.A., Treasurer and Hon. Librarian, Loughton. (2 copies.)
* Wallinger, R. N. A., Kitts Croft, Writtle, Chelmsford. (2 copies.)
* Walsingham, The Rt. Hon. the Lord, M.A., LL.D., F.R.S., F.L.S., F.E.S., Merton Hall, Thetford, Norfolk.
Watkins & Doncaster, 36, Strand, W.C.
* Wells, A. Godfrey, St. Alkmunds, Wanstead, N.E.
White, Chas., Holly House, Warrington.
White, Charles E., 57, North Hill, Colchester.
* White, J. H., Pease Hall, Springfield, Chelmsford. (2 copies.)
* White, William, F.E.S., The Ruskin Museum, Meersbrook Park, Sheffield.
* Whitaker, William, F.R.S., F.G.S., 3, Campden Road, Croydon.
* Whittle, F. C., 3, Marine Avenue, Southend, Essex.
Wicks, J., Dereham Place, Colchester.
Williams & Norgate, 14, Henrietta Street, Covent Garden, W.C.
* Wilson, T. Hay, Tudor Cottage, Clay Hill, Bushey, Herts.
* Wilson, The Rev. W. Linton, M.A., Rumburgh Vicarage, Halesworth, Suffolk.
* Winstone, B., F.C.S., F.G.S., F.R.M.S., 53, Russell Square, W.C.
* Wood, J. M., C.E., &c., 113, Balfour Road, Highbury New Park, N.
* Wright, Chas. A., F.L.S., F.Z.S., Knight of the Crown of Italy, Kayhough, Kew, Surrey.
Wright, Miss E. H., 7, Lexden Road, Colchester.

INDEX.

Abramis blicca, 24, 28, 112.
Abramis brama, 24, 25, 28, 111.
Acanthias vulgaris, 120.
Acerina vulgaris, 24, 88.
Acipenser sturio, 22, 24, 118.
Adder, 85.
Agonus cataphractus, 90.
Alburnus lucidus, 112.
Allis Shad, 22, 113.
Ammodytes lanceolatus, 101.
Ammodytes tobianus, 101.
Anarrhicas lupus, 95.
Anchovy, 112.
Angler, 91.
Anguilla vulgaris, 22, 24, 25, 26, 27, 28.
Anguis fragilis, 84.
Arvicola agrestis, 65.
Arvicola amphibius, 64.
Arvicola glareolus, 67.
Atherina presbyter, 96.
Azurine, 110.

Badger, 3, 40.
Balænoptera borealis, 78.
Balænoptera musculus, 78.
Ballan Wrasse, 98.
Barbel, 23, 24, 109.
Barbus vulgaris, 24, 109.
Bass, 88.
Bat, Barbastelle, 32 ; Common, 34 ; Daubenton's, 34 ; Greater Horseshoe, 31 ; Long-eared, 32 ; Mouse-coloured, 123 ; Noctule or Great, 34 ; Reddish-grey, 36 ; Serotine, 33 ; Whiskered, 36.
Beaumaris Shark, 119.
Beaver, 9.
Belone vulgaris, 106.
Bib, 99.
Black Fish, 92.
Bleak, 23, 112.
Blind Worm, 84.
Boar Fish, 93.
Bream, 24, 25, 28, 111.
Bream-flat, 24, 28, 112.
Brill, 20, 101.

Bubalis, 89.
Bufo vulgaris, 86.
Bull-head, Greenland, 89.
Butter Fish, 96.

Callionymus lyra, 94.
Canis mesomelas, 54.
Canis vulpes, 51.
Capreolus caprea, 77.
Capros aper, 93.
Carassius auratus, 109.
Carassius vulgaris, 108.
Carp, 24, 25, 26, 28, 108.
Carp, Crucian or Prussian, 108.
Centrolophus pompilus, 92.
Centronotus gunnellus, 96.
Cervus dama, 74.
Cervus elephas, 70.
Chubb, 23, 24, 26, 109.
Clupea alosa, 22, 113.
Clupea finta, 22, 114.
Clupea harengus, 113.
Clupea pilchardus, 113.
Clupea sprattus, 113.
Cod, 98 ; Large-headed, 99.
Conger, 115.
Conger vulgaris, 115.
Coregonus oxyrhynchus, 105.
Corkwing, 98.
Cottus bubalis, 89.
Cottus, Four-horned, 89.
Cottus gobio, 24, 26, 28, 89.
Cottus grœnlandicus, 89.
Cottus quadricornis, 89.
Cottus scorpius, 89.
Crenilabrus melops, 98.
Crossopus fodiens, 40.
Cyprinus carpio, 24, 25, 26, 28, 118.
Cystophora cristata, 56.
Cyclopterus lumpus, 95.

Dab, 102 ; Lemon, 102.
Dace, 24, 25, 26, 28, 110.
Deer, Fallow, 3, 74 ; Red, 3, 4, 70 ; Roe, 77.

Delphinus albirostris, 83.
Delphinus tursio, 82.
Dog Fish, Picked, 120.
Dolphin, 30 ; Bottle-nosed, 8, 82 ; White-beaked, 83.
Dormouse, 58.

Eel, 20, 22, 24, 28, 114.
Elleck, 90.
Engraulis encrasicholus, 112.
Erinaceus europæus, 37.
Esox lucius, 24, 25, 26, 28, 105.

Father Lasher, 89.
Fire Flaire, 121.
Fisheries, 13.
Fishing Frog, 91.
Flounder, 22, 24, 102.
Forkbeard, Lesser, 101.
Four-horned Cottus, 89.
Fox, 51.
Frog, 86.
Frog, Edible, 9, 86.

Gadus æglefinus, 99.
Gadus luscus, 99.
Gadus macrocephala, 99.
Gadus merlangus, 100.
Gadus morhua, 98.
Gadus pollachius, 100.
Galeus vulgaris, 119.
Garfish, 17, 106.
Gasterosteus aculeatus, 24, 25, 26, 28, 97.
Gasterosteus pungitius, 24, 25, 28, 97.
Gasterosteus spinachia, 98.
Gibbous Wrasse, 98.
Gobio fluviatilis, 24, 25, 26, 28, 109.
Gobius ruthensparri, 94.
Goby, One-spotted, 94 ; Two-spotted, 94 ; Slender, 94 ; Yellow, 94.
Gold Fish, 109.
Golsinny, 98.
Grampus, 81.
Grampus griseus, 81.
Grampus, Risso's, 81.
Grayling, 24, 28, 105.
Gudgeon, 24, 25, 26, 28, 109.
Gurnard, Grey, 90 ; Red, 90.

Haddock, 99.
Hake, 100.
Halichærus gryphus, 56.
Hare, 68.
Hedgehog, 37.
Herring, 113.
Hippocampus antiquorum, 117.
Horse-shoe Bat, Greater, 31.
Houting, 105.
Hyperoodon rostratus, 80.

Jackal, 54.
John Dory, 93.

Kettle-fishing, 18.

Labrax lupus, 88.
Labrus maculatus, 98.
Lacerta vivipara, 84.
Lamna cornubica, 117.
Lampern, 22, 24, 122.
Lamprey, Planer's, 122 ; Mud, 122.
Launce, Larger, 101 ; Lesser, 101.
Lepus cuniculus, 69.
Lepus timidus, 68.
Leuciscus cæruleus, 110.
Leuciscus cephalus, 24, 26, 109.
Leuciscus erythrophthalmus, 24, 27, 28, 110.
Leuciscus phoxinus, 24, 25, 26, 27, 28, 111.
Leuciscus rutilus, 24, 25, 26, 28, 109.
Leuciscus vulgaris, 24, 25, 26, 28, 110.
Ling, Three-bearded rock, 123 ; Five-bearded rock, 100.
Liparis montagui, 95.
Liparis vulgaris, 95.
Lizard, Viviparous, 84.
Loach, 24, 25, 26, 27, 28, 112 ; Spined, 28.
Lophius piscatorius, 91.
Lump Fish, 95.
Lutra vulgaris, 43.

Mackerel, 15, 18, 92.
Marten, 3, 49.
Meles taxus, 40.
Merluccius vulgaris, 100.
Miller's Thumb, 24, 25, 26, 28, 89.
Minnow, 24, 25, 26, 27, 28, 111.
Mole, 38.
Molge cristata, 87.
Molge palmata, 87.
Molge vulgaris, 87.
Monk Fish, 120.
Montagu's Sucker, 95.
Motella mustela, 100.
Motella tricirrata, 123.
Mouse, Common, 61 ; Harvest, 59 ; Wood, 60.
Mugil capito, 96.
Mullet, Grey, 96 ; Lesser Grey, 97.
Mus decumanus, 62.
Mus minutus, 59.
Mus musculus, 61.
Mus rattus, 61.
Mus sylvaticus, 60.
Mustela erminea, 47.
Mustela martes, 49.
Mustela putorius, 47.
Mustela vulgaris, 46.
Myoxus avellanarius, 58.

INDEX.

Natterjack Toad, 9.
Nemacheilus barbatula, 24, 25, 26, 27, 28, 112.
Nerophis æquoreus, 116.
Nerophis ophidion, 116.
Network Sucker, 95.
Newt, Common, 87; Palmated, 87; Water, 87.
Noctule, 34.
Orca gladiator, 81.
Orcynus thynnus, 92, 123.
Orthagoriscus truncatus, 117.
Osmerus eperlanus, 22, 105.
Otter, 43.
Oyster, 16.

Perca fluviatilis, 24, 25, 26, 28, 88.
Perch, 24, 25, 26, 28, 88.
Peter-netting, 17.
Petromyzon branchialis, 27, 122.
Petromyzon fluviatilis, 22, 24, 122.
Petromyzon marinus, 22, 27, 121.
Phoca barbata, 29.
Phoca fœtida, 29.
Phoca vitulina, 55.
Phocæna communis, 82.
Physeter macrocephalus, 79.
Pike, 24, 25, 26, 28, 105.
Pilchard, 113.
Pipe Fish, Broad-nosed, 116; Greater, 116; Ocean, 116; Straight-nosed, 117.
Piper, 90.
Plaice, 20, 101.
Planer's Lamprey, 26, 27, 122.
Plecotus auritus, 32.
Pleuronectes flesus, 22, 24, 102.
Pleuronectes limanda, 102.
Pleuronectes microcephalus, 102.
Pleuronectes platessa, 101.
Pogge, 90.
Polecat, 3, 47.
Porbeagle, 119.
Porpoise, 8, 82.

Rabbit, 69.
Raia alba, 120.
Raia batis, 120.
Raia clavata, 120.
Rana esculenta, 86.
Rana temporaria, 86.
Raniceps raninus, 101.
Rat, Hanoverian, 62; Black, 61; Water, 64.
Rhina squatina, 120.
Rhinolophus ferrum-equinum, 31.
Rhombus lævis, 101.
Rhombus maximus, 101.
Roach, 21, 24, 25, 26, 28, 109; Blue, 110. See Azurine.

Rorqual, 78.
Rorqual, Rudolphi's, 78.
Rudd, 21, 24, 27, 28, 110.
Ruff, 23, 24, 88.

Salmon, 20, 22, 23, 24, 26, 103.
Salmo fario, 24, 25, 26, 28, 104.
Salmo salar, 22, 24, 26, 103.
Salmo trutta, 22, 24, 26, 103.
Sand Eel, 101.
Sand Lizard, 9.
Sand Smelt, 96.
Sciurus vulgaris, 57.
Scomber scomber, 92.
Seal, 55; Bearded, 29; Common, 29; Grey, 56; Hooded, 56; Ringed, 29.
Sea Horse, 117.
Sea Lamprey, 22, 27, 121.
Sea Snail, 95.
Sea Trout, 22, 24, 26, 103.
Sea Wolf, 95.
Seine-netting, 17.
Shrew, Common, 39; Lesser, 39; Water 40.
Shrimp, 13, 20.
Silurus glanis, 27, 28, 106.
Skate, 120; Sharp-nosed, 120.
Slow Worm, 84.
Smelt, 17, 22, 105.
Snake, 8; Smooth, 9; Ringed, 84.
Sole, 20, 102; Lemon, 124; Little, 124; Red, 124; Sand, 124.
Solea lascaris, 124.
Solea lutea, 124.
Solea vulgaris, 102.
Solonette, 124.
Sorex minutus, 39.
Sorex vulgaris, 39.
Sprat, 16, 113.
Squirrel, 57.
Stag, 70.
Stickleback, Three-spined, 24, 25, 26, 28, 97; Ten-spined, 24, 25, 26, 28, 97; Fifteen-spined, 98.
Sting Ray, 121.
Stoat, 47.
Stowboats, 14, 16.
Sturgeon, 22, 24, 118.
Sun Fish, Short, 117; Oblong, 117.
Sweet William, 119.
Swift, Water, 87.
Sword Fish, 93.
Syngathus acus, 116.
Synotus barbastellus, 32.
Syphonostoma typhle, 116.

Talpa europæa, 38.
Tench, 24, 25, 26, 27, 28, 111.
Thornback, 120.
Thymallus vulgaris, 24, 105.

Tinca vulgaris, 24, 25, 26, 27, 28, 111.
Toad, 86.
Toad-fish, 91.
Toper, 119.
Trachinus draco, 91.
Trachinus vipera, 92.
Trawlers, 13, 16.
Trigla cuculus, 90.
Trigla gurnardus, 90.
Trigla hirundo, 90.
Trigla lyra, 90.
Tropidonotus natrix, 84.
Trout, 24, 25, 26, 27, 28, 104.
Trygon pastinaca, 121.
Tub Fish, 90.
Tunny, 92, 123.
Turbot, 18, 19, 101.
Twait Shad, 22, 114.
Vespertilio daubentonii, 34.
Vespertilio murinus, 123.
Vespertilio mystacinus, 36.
Vespertilio natterei, 36.
Vesperugo pipistrellus, 34
Vesperugo noctula, 34.

Vesperugo serotinus, 33.
Viper, 85.
Vipera berus, 85, 123.
Viviparous Blenny, 96.
Vole, Red Field or Bank, 67; Short-tailed Field, 65; Water, 64.

Weasel, 46.
Weever, Greater, 91; Viper, 92.
Well-boats, 14.
Wels, 27, 28, 106.
Whale, Common Beaked, 79; Sperm, 79.
Whitebait, 14, 20.
Whiting, 100.
Whiting Pollack, 100.
Whiting Pout, 99.
Wild Boar, 5.
Wolf, 5.
Wrasse, Ballan, 98; Gibbous, 98

Xiphius gladius, 93.

Zeus faber, 93.
Zoarces viviparus, 96.

THE ESSEX FIELD CLUB.
(Founded January 10th, 1880.)

Patron:
H.R.H. THE DUKE OF CONNAUGHT AND STRATHEARN, K.G.,
Ranger of Epping Forest.

President:
DAVID HOWARD, J.P., F.I.C., F.C.S.

Permanent Vice=Presidents:

PROFESSOR G. S. BOULGER, F.L.S. F.G.S.	T. V. HOLMES, F.G.S., M.A,I.
FREDERIC CHANCELLOR, J.P., F.R.I.B.A.	HENRY LAVER, M.R.C.S., F.L.S., F.S.A.
EDWARD A. FITCH, F.L.S., F.E.S.	PROFESSOR R. MELDOLA, F.R.S., F.R.A.S.

The Essex Field Club is intended to band together those taking an interest in Natural Science, residing within or near the borders of the County, so as to create and foster a taste for the out-of-door study of Nature.

The PUBLICATIONS of the Club have gained general estimation, because of the care which has been taken to confine them to their original purpose, the record of investigations and elucidations of the NATURAL HISTORY (in its widest sense), the TOPOGRAPHY and PRE-HISTORIC ARCHÆOLOGY of the COUNTY OF ESSEX. Since the foundation of the Club in 1880, over 4,500 pages of such material has been published, and a large proportion of the articles are of value to students residing outside the Club's limits.

ORDINARY (SCIENTIFIC) MEETINGS are held at frequent intervals for the reading of papers and the exhibition of specimens, etc., and FIELD MEETINGS are arranged during the summer months, and held in various parts of the County under the guidance of experienced Naturalists, Geologists, and Archæologists.

The Club has carried on several SPECIAL INVESTIGATIONS—*e.g.*, the examination of the two EPPING FOREST CAMPS, an EXPLORATION of the DENEHOLES, the "RED-HILLS," etc, and has published several valuable REPORTS, illustrated with plans and maps. Considerable efforts have been made to catalogue the FAUNA and FLORA of the County, and it is wished to extend this work as funds and opportunities will permit, particularly in the direction of a systematic Exploration for MARINE and ESTUARINE forms of life by DREDGING, etc., and their preparation and preservation in the Essex Museum, for future reference and study.

Although bearing a county title, the Club offers exceptional advantages to metropolitan residents. Many parts of Essex are but little known, although of the greatest interest to the naturalist, geologist, and antiquary.

Very considerable material has been accumulated towards a LOCAL and EDUCATIONAL MUSEUM, which will shortly be established in a handsome building at Stratford (by the generosity of Mr. Passmore Edwards, and in conjuction with the corporation of West Ham) to form a home for County collections and specimens, where they may be consulted by all interested in Essex A BRANCH MUSEUM to illustrate the Natural History and Archæology of EPPING FOREST has been established (under the sanction of the Corporation of London) in QUEEN ELIZABETH'S LODGE, CHINGFORD, which has proved very attractive to thousands of visitors to the Forest.

The Club already possesses a good nucleus of a Local and Scientific LIBRARY, which has been obtained by donations, exchanges, and purchase.

The *Minimum* SUBSCRIPTION is Fifteen Shillings per annum, payable upon election, and afterwards on the 1st January in each year. The usual entrance fee is at present in abeyance. The LIFE CONPOSITION is £10 10s. in one payment. *Members can purchase the publications of the Club at a Discount of 25 per cent. from the published prices.*

Copies of the RULES, FORMS of PROPOSAL for MEMBERSHIP, together with specimen copies of the ESSEX NATURALIST, and other information, will be gladly sent on application to the *Hon. Secretaries*, Messrs. W. and B. G. COLE, Buckhurst Hill, Essex.

PUBLICATIONS.

The *Transactions* and *Proceedings of the Essex Field Club* (1880-1886) record tne first six years work of the Society. The complete set of five volumes (unbound, about 1200 pp.) is £2 16s. 0d.

The *Essex Naturalist* was commenced in 1887; Vol. X. is now in course of publication. The set of the completed nine volumes (unbound) is £4 4s. 0d.

SPECIAL MEMOIRS. *VOL. I. Report on the East Anglian Earthquake of April 22nd,* 1884, by Prof. R. Meldola, F.R.S., and William White (bound), 3s. 6d.

VOLUME II. The Birds of Essex, a contribution to the Natural History of the County, by Miller Christy, F.L.S. (bound), 10s. 6d. (to members).

VOLUME III. The Mammals, Reptiles, and Fishes of Essex, by Dr. Henry Laver, F.L.S., to members (by subscription only), 6s., neatly bound.

THE ESSEX NATURALIST.

(*The Journal of the Essex Field Club.*)
Edited by WILLIAM COLE, F.L.S., F.E.S.

Published Quarterly. Free to Members; to Non-Members price 6s. per annum. Post free.

The *Essex Naturalist* contains PAPERS and NOTES ON NATURAL HISTORY, GEOLOGY, and PRE-HISTORIC ARCHÆOLOGY, having special reference to Essex, or interesting to Field Naturalists and Geologists in the LONDON DISTRICT and frequenters of EPPING FOREST, as well as hints and aids on the study of NEGLECTED GROUPS of Animals and Plants occurring in the district.

COMMUNICATIONS *(Particularly Notes of Observations)* on the above-named class of subjects will be welcomed.

ADDRESS:—Editor, *Essex Naturalist,* 7, Knighton Villas, Buckhurst Hill, Essex.

A SELECTION FROM

MESSRS. EDMUND DURRANT & CO.'S
LIST OF PUBLICATIONS.

A Chart showing the History and Formation of the Diocese of St. Albans. By the Rev. D. W. BARRETT. On thick paper, 1s.; by post, 1s. 3d.

Marmaduke, Emperor of Europe. By X. Crown 8vo, illustrated, 6s., cloth.

Household Hints to Young Housewives. By MARTHA CAREFUL, Sewed, 6d.

The Holy City: Jerusalem, Its Topography, Walls and Temples. A New Light on an Old Subject. By Dr. S. R. FORBES, author of "Rambles in Rome," etc. Crown 8vo. cloth, illustrated, 2s. nett.

The Essex Review. Edited by E. A. FITCH, F.L.S., and Miss FELL SMITH, published quarterly. Annual Subscription, 5s., post free; single Nos., 1s. 6d. nett. (See advert. hereafter.)

Some of our East Coast Towns. By J. EWING RITCHIE. Crown 8vo, sewed, 6d.

The Ancient Sepulchral Monuments of Essex. By FRED CHANCELLOR, F.R.I.B.A. Imp. 4to, cloth, illustrated, £4 4s. nett.

Poems. By ALICE E. ARGENT, with an introduction by the Right Rev. BISHOP CLAUGHTON. Crown 8vo, cloth, 3s. 6d. nett, post free.

Durrant's Handbook for Essex. A Guide to all the Principal Objects of Interest in each parish in the County. By MILLER CHRISTY, F.L.S. With Maps, 2s. 6d. nett, post free. (See advert. hereafter.)
"One of the best guide books in existence."—*Evening News*.

The Birds of Essex. A contribution to the Natural History of the County. With numerous illustrations, two plans, and one plate (forming Vol. II. Special Memoirs of Essex Field Club). By MILLER CHRISTY. Demy 8vo. scarlet cloth, 10s. nett, post free. (See advert. hereafter.)

A History of Felstead School. With some account of the Founder and his Descendants. By JOHN SARGEAUNT, M.A. Illustrated, nett 4s.

The Trade Signs of Essex. A popular account of the origin and meaning of the Public House and other signs now or formerly found in the County of Essex. With illustrations. By MILLER CHRISTY. Demy 8vo, cloth, 7s. 6d. nett.

Daily Rays of Light for Sick and Weary Ones. Compiled by EDITH L. WELLS, with a preface by the Rev. PREBENDARY HUTTON. Crown 8vo, cloth, 6s.

The Limits of Ritual in the Church of England. By the Rev. R. E. BARTLETT, M.A., late Fellow and Tutor Trinity College, Oxford; Bampton Lecturer, 1888. Reprinted by permission from "CONTEMPORARY REVIEW." Price 3d., by post 3½d.; 2s. 9d. per dozen, post free.

Homespun Yarns. By EDWIN COLLER. Crown 8vo, 3s. 6d.

Royal Illustrated History of Eastern England. By A. D. BAYNE, with many illustrations. Two volumes, large 8vo, cloth, 15s.

Domesday Book relating to Essex. Translated by the late T. C. CHISENHALE-MARSH. 4to, cloth, 21s. nett. Only a few copies unsold.

John Nokes and Mary Styles. A poem in the Essex Dialect. By the late CHARLES CLARK, of Totham Hall. With a Glossary and Portrait, 1s. nett.

The History of Rochford Hundred, Essex. Vol. I., 15s. 6d.; Vol. II., 18s. nett. By PHILLIP BENTON.

A First Catechism of Botany. By JOHN GIBBS. Second Edition, 12mo, boards, 6d.

The Symmetry of Flowers. By JOHN GIBBS. 18mo, sewed, 4d.

Forms and Services used in the Diocese of St. Alban's. Published by authority. *Lists on application.*

EDMUND DURRANT & CO., 90, High Street, CHELMSFORD.

Demy 8vo, Red Cloth, 10s. nett.

THE BIRDS OF ESSEX.

A CONTRIBUTION

TO THE

NATURAL HISTORY OF THE COUNTY.

By MILLER CHRISTY, F.L.S.

With 162 Woodcut Illustrations, 2 Plans, and a Frontispiece.

PRESS NOTICES.

"This work . . . does equal credit to the enterprise of the Essex Field Club and the author. . . . The work is thoroughly well done, and is a valuable addition to our local lists."—*Athenæum.*

"Mr. Christy's volume, with its valuable original contributions from local ornithologists, will be a permanent addition to British zoology, and a stimulus to the Field Clubs throughout England."—*Leisure Hour.*

"This book, profusely illustrated, places a rich fund of information at the disposal of all who love our feathered friends, collected by one who evidently knows them well."—*Pearson's Weekly.*

"Mr. Christy makes here an interesting contribution to the natural history of his county."—*Spectator.*

"Almost every county nowadays has its local bird-chronicle, and it seems remarkable that a county ornithologically so rich as Essex should have remained a field unattempted until Mr. Christy took it up. . . . His book is full of information excellently put together."—*Saturday Review.*

"Mr. Miller Christy's recently published and excellent work on the 'Birds of Essex' deserves the attention of every rambler in Epping Forest."—*The Globe.*

"This work . . . differs from most works of the kind in being profusely illustrated. That most of the figures of birds are not new is, in this instance, no drawback, for they have been well chosen."—*The Field.*

"In dealing with the mass of material that has come to hand [Mr. Christy's] industry and judgment are on every page."—*Zoologist.*

"We have now before us a volume on the Birds of Essex, for which, the author tells us, he has been collecting information and materials for fifteen years. It seems to fully correspond in completeness to the time spent upon it. . . . Mr. Christy's volume is, in fact, well planned and well got up, and it has the great merit of not being too bulky."—*Ibis.*

"The 'Birds of Essex' . . . is a valuable addition to the literature of the county. . . . The work bears evidence of great industry and literary skill."—*Essex County Chronicle.*

CHELMSFORD :
EDMUND DURRANT & CO., 90, High Street.

LONDON :
SIMPKIN, MARSHALL & CO., LTD.

Fcp. 8vo.   2s. 6d. nett.   Post Free.

DURRANT'S HANDBOOK FOR ESSEX.
A GUIDE TO
The Principal Objects of Interest in each Parish in the County, for the use of Tourists and others, with an Introduction treating of its History, Geology, Area, Population, Dialect, Antiquities, Worthies, Natural History, etc., etc.

By MILLER CHRISTY, F.L.S.
With Maps.

PRESS NOTICES.

"... That well-known lover of and writer on Essex, Mr. Miller Christy, treats of its history, geology, population, literature, antiquities, natural history, and last, but not least, its long list of 'worthies.' Some notes on the dialect will interets the scholar, and the general information appears accurate and trustworthy."—*Daily Telegraph.*

"This little book might be commended as a model to writers of Guide Books. Its whole arrangement represents intelligent care and thought, and it is doubtful whether there could be found in any similar work such a mass of useful information in so singularly small a compass."—*Athenæum.*

"Very well arranged. In alphabetical order. No Essex man should be without a copy."—*Literary World.*

"The quantity of valuable information comprised within its limits is marvellous."—*Essex Weekly News.*

"No naturalist intending to visit Essex should fail to provide himself with a copy."—*Naturalists' Monthly.*

"A really valuable guide to the places of interest in the county, a work which was much ueeded and will be much appreciated."—*Essex Standard.*

"A handy volume for Essex wheelmen; full of very interesting historical and general information concerning the various towns and villages in Essex."—*Wheeling.*

CHELMSFORD : EDMUND DURRANT & CO., 90, HIGH STREET.
LONDON : SIMPKIN, MARSHALL & CO., STATIONERS' HALL COURT.

THE ESSEX REVIEW.
An Illustrated Quarterly Record of
Everything of Permanent Interest in the County of Essex.
Edited by E. A. FITCH, F.L.S., and MISS C. FELL SMITH.

"A local organ of unusual excellence. Its appearance does great credit to Messrs. Durrant's Chelmsford press, and its contents are above the average of such productions . . . If kept up to its present standard, the REVIEW will be widely appreciated."—*Athenæum.*

"This ably conducted and interesting quarterly." "The REVIEW is decidedly creditable to the energy and local patriotism of the projectors."—*Daily News.*

"Biography, history, literature, &c., of Essex treated in a scholarly fashion."—*Daily Chronicle.*

"We have praised the ESSEX REVIEW, because, on the whole, the work in it has really been worthy of praise, and because we think it is eminently useful that the more important facts pertaining both to the past and present, should be embalmed in a volume which can be easily referred to, and relied upon with confidence."—*Essex County Chronicle.*

"The magazine well maintains the excellent reputation it has gained of being bright and readable."—*Essex Weekly News.*

"The REVIEW deserves the support of Essex people . . . and has the recommendation of treating of news modern as well as of news ancient."—*Essex Standard.*

CHELMSFORD : EDMUND DURRANT & Co.
LONDON : SIMPKIN, MARSHALL & Co., 4, STATIONERS' HALL COURT.
Quarterly : 1/6 Nett.   Or 5/- Yearly, Post Free, if paid in advance.
Send for a Specimen Copy, 1/6.

www.ingramcontent.com/pod-product-compliance
Lightning Source LLC
Chambersburg PA
CBHW030248170426
43202CB00009B/665